WHERE DO CAMELS BELONG?

KEN THOMPSON

WHERE DO CAMELS BELONG?

*Why Invasive Species
Aren't All Bad*

GREYSTONE BOOKS

Vancouver/Berkeley/London

Greystone Books Ltd.
greystonebooks.com

Cataloguing data available from Library and Archives Canada
ISBN 978-1-77164-096-1 (pbk.)
ISBN 978-1-77164-097-8 (epub)

Cover design by Peter Dyer
Text design by Henry Iles
Cover illustration by Getty Images

Greystone Books thanks the Canada Council for the Arts,
the British Columbia Arts Council, the Province of British
Columbia through the Book Publishing Tax Credit, and the
Government of Canada for supporting our publishing activities.

Canada

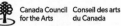

Greystone Books gratefully acknowledges the xʷməθkʷəy̓əm (Musqueam),
Sḵwx̱wú7mesh (Squamish), and səlilwətaɬ (Tsleil-Waututh) peoples on
whose land our Vancouver head office is located.

CONTENTS

CONTENTS

WHERE DO CAMELS BELONG?

Where do camels belong? Ask the question and you may instinctively think of the Middle East, picturing a one-humped dromedary, some sand and perhaps a pyramid or two in the background. Or if you know your camels and imagined a two-humped Bactrian, you might plump for India and central Asia. But things aren't quite so simple if we're talking about the entire camel family.

Camelids (the camel family) evolved in North America about 40 million years ago. *Titanotylopus*, the largest camel that has ever lived, stood 3.5 m high at the shoulder and ranged through Texas, Kansas, Nebraska and Arizona for around 10 million years. Other species evolved very long necks and probably browsed on trees and tall shrubs, rather as giraffes do today. Much, much later camels spread to South America, and to Asia via the Bering

Strait, which has been dry land at various times during the recent Pleistocene glaciations. Camels continued to inhabit North America until very recently, the last ones going extinct only about 8,000 years ago. Their modern Asian descendants are the dromedary of north Africa and south-west Asia and the Bactrian camel of central Asia. Their South American descendants are the closely related llamas, alpacas, guanacos and vicuñas (llamas are only camels without humps; all you need to do is look one in the eye for this to be pretty obvious).

Now you know all that, let me ask you again: where do camels belong? Is it:

(a) in the first place you think of when you hear the word 'camel', i.e. the Middle East.

(b) in North America, where they first evolved, lived for tens of millions of years, achieved their greatest diversity, and where they became extinct only recently.

(c) in South America, where they retain their greatest diversity.

Or, just to muddy the waters a bit more, is it:

(d) in Australia, where the world's only truly wild (as opposed to domesticated) dromedaries now occur.

Finally, if you felt able to give a confident answer, can you explain why?

If you think camels belong where they evolved, the question has only one answer: North America. If it means where they have been present for the longest time, the answer is the same. If it means where camels have been present during recent millennia, then the answer is Asia and South America. If camels belong wherever they can thrive without human assistance, then it must also include Australia. These are all perfectly reasonable interpretations of belonging.

And there is nothing particularly special about camels. Dispersal over huge distances is not at all unusual among land animals, and it is almost routine among birds. Horses are much the same as camels, and frogs, toads, shrews, deer, cats, weasels, otters, hares, skinks, chameleons and geckos are among the many other groups that now occur almost everywhere, and do so as a result of relatively recent dispersal – without human assistance – often starting out in Africa or south-east Asia. None of these species have an obvious answer to the question about where they belong – whether they are natives or aliens – any more than camels. Indeed, once you adopt a view of the world that doesn't assume that there's something very special about where things happen to be right now (or in relatively recent history), asking where *anything* belongs tends not to have an obvious answer.

⬤ ⬤ ⬤

The Earth is home to just short of two million species of living organisms. At least, those are the ones we have recognised, described and named. There are certainly many more, maybe up to 10 million, possibly even more. Each of those species has a characteristic distribution on the Earth's land surface, or in its oceans, lakes and rivers. Some are common, some are rare, some have very wide ranges, others are confined to tiny areas such as single islands. But in every case, that distribution is in practice a single frame from a very long movie. Run the clock back only 10,000 years, less than a blink of an eye in geological time, and nearly all of those distributions would be different, in many cases very different. Go back only 10 million years, still a tiny fraction of the history of life on Earth, and any comparison with present-day distributions becomes impossible, since most of the species themselves would no longer be the same. Go back further still, and the Earth itself starts to become unfamiliar, with some continents drifting further apart, others colliding.

Only rarely do we get a really good view of what a dynamic, unstable place the world and its inhabitants really is, but when we do it can be quite startling. Recently, Dutch researchers drilled down over half a kilometre to obtain sediment cores from the Bogotá basin in the tropical High Andes of Colombia. The pollen grains preserved in this sediment column tell us what the vegetation was like at every moment during the last two million years – and the researchers found something remarkable. This is what they concluded:

> *Present-day montane forest and páramo vegetation reflect a 'frozen moment' in a long and dynamic process of almost continuous reorganization of floristic elements. It indicates that on a Pleistocene timescale present-day plant associations are ephemeral. Most of the record reflects no-analogue vegetation associations.*

In other words the plants (and the vegetation they formed) that would have been familiar to a human observer at any moment during the last two million years would have seemed quite unfamiliar to anyone from any other point in time. Not only that, but (that final 'no-analogue' comment) none of the various kinds of vegetation that grew during that immense span of time has any close modern equivalent, and all would be unfamiliar to a present-day observer.

What all this tells us is that there is nothing special about the plants – or camels or anything else – we have now, nor about exactly where they happen to be, i.e. where they are currently 'native'. The only unusual thing about *now* is that we are here to see it. Which, of course, prompts another question. If we consider the Colombian example above, is there any sense in which any of the different kinds of vegetation that have existed there are *better* – or worse – than what we have now? Does the long-vanished flora of, say, a million years ago have any more *right* to occupy the Bogotá basin than the vegetation that was around two million years ago, or than what is there today?

If we believe the answer to either or both of those questions is yes, then we need to answer another question: which vegetation do we prefer? If there is a hierarchy of rights and belonging, who or what is at the top? And why? And, most urgently of all, how can we reply to that question in such a way that the answer is given a good, shiny coat of scientific objectivity?

One answer is to observe that man is now by far the most important disperser of species around the globe, and to assert that human interference with species' distributions is an unnatural process – in effect that mankind is now no longer part of the natural world. Essentially that man is now bringing together species that, without our intervention, would have taken a very long time to meet, or might never have met. Yet if a study of the history of life on Earth teaches us anything, it's that 'never' should be used with extreme caution. The unique mammals of South America, evolving in isolation for 100 million years, must have thought they would never encounter their more advanced cousins from North America – until they did.

If we subscribe to this view of the world, we do not need to know *why* dispersal of species by humans is inherently unnatural. Nor do we need to know what event – whether the invention of agriculture, or the steam engine, or the lawnmower – caused *Homo sapiens* to be forever sundered from the rest of creation. It is enough to know that just before this event the Earth's species were briefly, and for the first, last and only time, not only where they ought to be, but also where they ought to remain. (Nor, apparently, is this invalidated in any way by the massive human modification of the majority of the Earth's surface, rendering much of it quite unsuitable

5

for the species that used to live there, nor by current and future anthropogenic climate change, which threatens even those few parts of the globe that remain relatively untouched by man.)

If we adopt this idea – and, bizarre though it seems, it has become a dominant and orthodox view – the 'frozen moment' when there was a place for everything, and everything knew its place, is set not quite now but at some point in the pre-human, pre-industrial past. Everything and anything that has happened since (which by definition would have turned out very differently without human intervention) is wrong in practice and in principle. And in such manner, belonging – or 'nativeness' – is elevated into one of the great conservation principles of our time, conferring indefinite rights of future occupancy and significant public funding on species judged to possess that nebulous quality, and zealous persecution of those species deemed not to belong.

This black-and-white view of the world – 'natives' good, 'aliens' bad – is justified by a focus on a relatively few species that cause undoubted economic or environmental harm when moved to new areas. But it ignores the vast majority that do no harm at all, or are positively useful – including practically all the crop plants and animals on which human civilisation depends. It is also based on multiple distortions in defining 'nativeness'. Adopting the frozen moment as one's perspective leads into the temptation to regard attractive, harmless (and especially rare) species as native; and, conversely, to consider species we don't like as alien. We rather too easily attach the pejorative epithet 'invasive' to 'alien', so that before you know it all aliens are 'invasive aliens'. And even if they're not obviously invasive (whatever that means), we suspect that one day they will be, or that we haven't looked hard enough for evidence of their delinquency.

Of course, native species often move around too, but such movements, whatever their impacts, are not considered 'invasions'. Indeed, even the movements of aliens that we've decided we like – such as, in Britain, the recent spread of little egrets into the south of England – are tagged as 'migrations'. The rest of the vocabulary of biological invasions is similarly elastic: once we agree that alien species are by definition harmful, their presence itself becomes one measure of 'harm', and because we 'know' that alien species cause economic damage we routinely inflate the cost of such damage by ignoring any possible beneficial impacts.

⸙ ⸙ ⸙

You might by this point be wondering whether I'm just paranoid. There surely isn't a global conspiracy to promote and maintain such a view of the world? Well, yes and no. There is no conspiracy, but a remarkable coalition has developed to promote this version of reality.

For biologists, alien species provide unparalleled opportunities to study dispersal, colonisation, competition and evolution in action. But funding for such pure research is limited, so there's an understandable tendency to loosen the purse strings by presenting aliens as some kind of existential threat to life. Not just species currently judged to be invasive either, but also those with the potential, however remote, to become invasive. Conservationists are, too often, happy to go along with this, because conservation is a value-laden activity, whose values are not always easy to pin down. 'Nativeness' appears to offer the prospect of unambiguous attributes that make something worth conserving; or, in the case of its absence, worth exterminating, or at least controlling. To question this approach is close to heresy. And the media are happy to buy into it. The language is easy to put across: the respect for natives and (especially) the fear of aliens.

7

This book is an examination of the whole question of native and alien species, and what might almost be called an alien invasions industry – and its implications. Along the way, I do my best to answer many awkward questions. Should we worry about alien species and, if so, how much should we worry? How much truth is there in the 'alien invasion' scare stories we encounter every day? Is there a genuine 'cost' to these invasions? How well do we really understand the biology of invasions by alien species? And indeed is there any fundamental distinction between such invasions and the normal ebb and flow of native species? Only a minority of introduced species succeed, so is there anything special about that minority, and in particular about the even smaller minority that go on to cause any trouble? Do we always know what's native and what isn't, and what do we really mean by native anyway? How good is our record of controlling or eradicating alien species, and do we always choose the right targets? Do exotic species always cause as much trouble as we think they do, or are most just making the best of the mess left by the most dangerous species of all – *Homo sapiens*? And is fear of invasive species getting in the way of conserving biodiversity, and especially of responding to the threat of climate change?

So if you've ever read a newspaper headline about the invasion of the alien swamp monster, or the arrival of Japanese knotweed in your local park, and wondered what all the fuss is about, this is the book for you.

CHAPTER ONE

SPECIES ON THE MOVE

S **pecies are born, and then they die.** That is, they evolve
by natural selection from earlier species, and eventually
they go extinct. In the intervening period, which may
be less than a million years, or as much as ten million years (or
even more), they may do many things. But one thing they don't
do is hang around in the same place. More or less by definition
species evolve in one particular spot, but later they may spread
to occupy a much wider range, a range that often does not
include their original 'home'. The eventual range may be large
or small, it may be continuous or divided into smaller patches.
The variety can be bewildering. Indeed, biologists, palaeontolo-
gists and geographers have grappled for the past hundred-or-so
years with the problem of explaining how species come to be
where they are and (sometimes an even harder question) not
where they are not.

Early biogeographers were baffled by some animal and
plant distributions because the history of the Earth's continents

was a mystery to them. Why are marsupial mammals and the Southern beech (*Nothofagus*) found in South America and Australia, and both also as fossils in Antarctica? Alfred Wegener proposed his theory of continental drift in 1912, but his ideas (now known as plate tectonics) were not widely accepted until the late 1950s.

SPECIES AND CONTINENTS

We now know that South America, Australia, Antarctica and Africa were once united into a single supercontinent, Gondwana, and practically all of the biogeography of the southern hemisphere can be explained by the timing of its break-up. India broke away first, soon followed by Africa, so neither acquired the early marsupials that spread throughout the rest of Gondwana. But Africa did remain attached for long enough to acquire the large, flightless ratite birds (African ostrich, South American rheas, Australian cassowaries and New Zealand kiwis and extinct moas), plus some characteristic southern hemisphere plants, such as Gunnera, the odd, rush-like restios and the Proteaceae; *Protea cynaroides* (king protea) ultimately going on to become South Africa's national flower.

The history of the northern hemisphere has been quite different, and also quite complicated, but the basic fact to remember is that the northern hemisphere continents have generally been more joined up, and until more recently, than the southern. In fact the waters of the Bering Strait are relatively shallow and have been repeatedly exposed as dry land when sea level has been lower during the several recent glacial episodes. These connections have allowed animals and plants that evolved in Asia to spread to North America and vice versa. As a result, there is a certain uniformity about northern

10

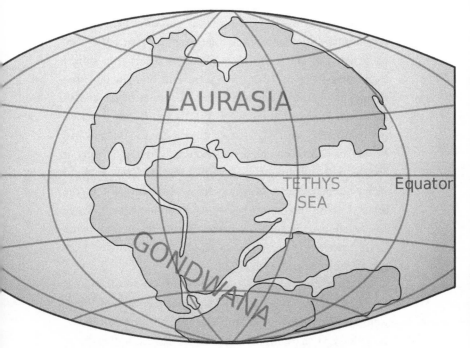

Pangaea, which contained all the world's land, began to break up around 200 million years ago into northern (Laurasia) and southern (Gondwana) supercontinents.

hemisphere floras and faunas. Indeed there's a large group of plants that are distributed more or less right round the globe at Arctic or cool temperate latitudes. Familiar examples, all found on hills or mountains in the UK, are mountain avens, moss campion and cowberry. Few mammal species do exactly the same (the brown bear is an exception), but in almost every case the North American and Eurasian species are either very closely related species, or sometimes only subspecies, for example, bison, red fox, otter, beaver and lynx. Sometimes we give them different names, such as moose and caribou in North America and elk and reindeer respectively in Eurasia, but they're still essentially the same.

11

Throughout most of prehistory, connections between the northern and southern hemispheres were few, but when they occurred they led to some of the most spectacular animal migrations the world has ever seen. The best documented came with the (relatively recent) establishment of the central American land bridge between North and South America. Ever since the break-up of Gondwana, South America had existed in almost complete isolation, developing a fauna at least as odd and unique as that of Australia. A few intrepid island-hoppers made the jump from one continent to the other (in both directions) before the continuous land bridge developed, but the main exchange, usually referred to as the Great American Interchange, occurred only three million years ago, towards the end of the Pliocene.

At first, large numbers of different kinds of animals migrated in both directions, and some of the southern kinds were initially highly successful. The ground sloth *Megalonyx*, weighing up to a tonne, made it as far as Alaska. South America did not have any advanced mammalian carnivores, so many of its top predators were birds, including the so-called 'terror birds', large flightless carnivores. Terror birds stood up to three metres tall and had the largest bird skulls known, with huge curved beaks; they were also fast runners, perhaps achieving speeds of close to 50 km an hour. One of the larger species spread as far north as Florida and Texas. Ultimately, however, all the larger southern colonists became extinct, and the only major mark the Interchange has left on the mammal fauna of North America is porcupines, anteaters, opossums and arma-dillos. Some smaller birds were also highly successful in the north: hummingbirds started out in South America but now extend as far north as Canada.

In contrast, most North American colonists of the south were ultimately far more successful, including canids (dogs, wolves and foxes), llamas, horses and (especially) rodents; nowadays, 85 per cent of modern South American mammals are of North American origin. We began this book by talking about camels, but horses show remarkable parallels. Again they arose in North America; 50 million years ago, the earliest horse ancestors were small, forest-dwelling browsers. Horse evolution is closely linked to changing climate and the spread of grass-lands across North America, as over time they gradually became larger and leggier, with tougher teeth, adaptations to running in open grasslands and eating grasses. North American horses were diverse, with several coexisting species. Just like camelids, horses spread to the Old World and also to South America, before becoming extinct in North America about 8,000 years ago. But in the case of the horse, the South American species, a pony-sized animal that probably looked a bit like a donkey, also became extinct at about the same time.

Charles Darwin, while on the *Beagle* expedition, famously found a fossil horse tooth in Patagonia, thus becoming the first person to realise that horses had existed in the Americas long before they were (re)introduced by the Spanish. The discovery had a powerful effect on Darwin's ideas about extinction, as he wrote in *The Origin of Species*:

> No one I think can have marvelled more at the extinction of species, than I have done. When I found in La Plata the tooth of a horse embedded with the remains of Mastodon, Megatherium, Toxodon, and other extinct monsters, which all co-existed with still living shells at a very late geological period, I was filled with astonishment; for seeing that the horse, since its introduction by the Spaniards into South America, has run wild over the whole country and has increased in numbers at an unparalleled rate, I asked myself what could so recently have exterminated the former horse under conditions of life apparently so favourable.

RELICTS, REFUGIA AND ICE AGES

It's clear that the movement of continents, and the migration of species across and between continents, all against a background of evolution of new species and extinction of older ones, have drastically rearranged the Earth's biota (its total collection of organisms) on numerous occasions. Some gaps in distributions, sometimes very large ones, are the result of dispersal, while others arise from the extinction of intervening populations by changes in climate. Nile crocodiles are (or were) ubiquitous from South Africa to the Nile Delta and across to Senegal. But what are they doing in the Tibesti Mountains in northern Chad, one of the most remote places on Earth and surrounded in every direction by hundreds of miles of Saharan desert?

Nile crocodiles – ubiquitous across a huge swathe of Africa.

Migration across that desert is clearly impossible for a wetland species, so the only possible explanation is that the climate was once much wetter, allowing crocodiles to extend much further north and west of their present continuous distribution. Plenty of other evidence points in the same direction, including geomorphological signs of former rivers and cave paintings of elephants and giraffe (both now absent) in the Tibesti. In fact we now know that the Sahara has rarely been as dry as it is right now.

Similar climatic relicts turn up almost anywhere you care to look. Crocodiles are an example of a wide contiguous distribution with a few odd outliers, but often an entire distribution will consist of isolated patches, with no obvious large centre of distribution. *Rhododendron ponticum*, familiar now to gardeners, is an example of a species with a formerly wide distribution in previous interglacial warm periods, but now reduced to a few isolated remnants. Its largest native territory is in the Pontic region, around the southern and eastern shores of the Black Sea, but it also occurs in western Turkey, in a few tiny areas of western Spain and Portugal and in the mountains of Lebanon. Alpine species, widely distributed in cold periods but restricted to mountains in warmer times, are particularly likely to show relict distributions. For example *Rhododendron ferrugineum* (alpenrose), another of Europe's few surviving rhododendrons, now survives only in the Alps, Pyrenees, Jura, northern Apennines and the Dinaric Alps in Croatia.

Nowhere is the ebb and flow of species quite so clear, or so well documented, as in the expansion and contraction of species' ranges in response to the Pleistocene glaciations in Europe and North America. Throughout the whole history of the Earth, glaciations ('ice ages') have been relatively rare, but the whole of the Pleistocene (the last 2.58 million years) has technically been an 'ice age', with a permanent ice sheet throughout that time in Antarctica and most likely in Greenland too. Periodically the ice sheets have expanded (glaciations), extending as far south as London, Seattle and New York at their greatest extent, and contracted (interglacials). We are currently about 10,000 years into an interglacial. These huge fluctuations in temperature have caused warmth-loving species to retreat to southern refugia during the glaciations, and then rapidly expand their ranges during the warm periods. These retreats and advances demonstrate both how fast species can migrate when conditions allow, and also how much chance is involved in how far they manage to advance, exactly where they retreat to, and indeed whether they manage to survive at all.

The British Isles are particularly interesting, because sea levels fall dramatically (by about 120 m) during a glaciation, creating a land connection between Ireland, Britain and mainland Europe. As the ice melts, Ireland and Britain once again become islands, Ireland significantly earlier owing to the greater depth of the Irish Sea. So there are always species that manage to reach Britain as the climate warms, but fail to reach Ireland, which is why (with apologies to St Patrick) Ireland has no snakes in the present interglacial. Or moles or common toads, come to that. However, every interglacial is different. In our current one (the Holocene, which has persisted for the past 11,700 years), silver fir (*Abies alba*) never made it to Britain, while in the previous one it got all the way to the west coast of Ireland. Similarly, box (*Buxus sempervirens*) managed to reach Ireland in the last interglacial, but in this one it only just made it to southern England.

But where did these species survive during the cold periods? This is where the story gets really interesting, because the geography of southern Europe means that Iberia, Italy and the Balkans provide alternative warm refuges. The oaks, shrews, water voles and (now extinct) bears that live in Britain in the present interglacial spent the glaciation in Spain, while the tawny owls, grasshoppers, newts and alder and beech trees reinvaded from the Balkans. The Alps make the use of Italy as a refuge somewhat problematical, but German and Scandinavian hedgehogs managed to make it back from there, while British hedgehogs came from Spain.

The implications of all this are profound. A crucial thing to remember is that, although we talk of animals and plants advancing and retreating as the glaciers waned and waxed, only the first part of that is literally true. Although species did spread north as the ice melted, those northern colonists generally died out when the ice returned. That is, there was no actual 'retreat', and the only populations that survived the whole cycle were those that remained in the southern refugia, mostly adjusting to rising and falling temperatures by ascending and descending the nearest mountain. So, although the English are fond of referring to their oaks as 'English oaks', the oaks that thrive in England today spent more than 99 per cent of the last two and a half million years in Iberia and are more accurately 'Spanish'. Not only that, but the refugium that provided specific colonists seems to be rather random, so English oaks in a previous inter-glacial may well have been Greek, or perhaps Italian.

One consequence of the repeated opening and closing of the English Channel is that there were always species that were capable of living in Britain but nevertheless failed to migrate fast enough and so were left behind in mainland Europe, and every now and then one of these failed colonists finally arrives. For example several bumblebees are found in Europe, but not in Britain, and one that many naturalists always thought 'should' be here is the tree bumblebee, *Bombus hypnorum*. The tree bumblebee is found throughout much of Europe as far north as the Arctic Circle, and it also seems to like gardens, so it generally became more abundant during the twentieth century. So it was no great surprise among the bee cognoscenti when a specimen was captured in southern England in 2001. Since then, the tree bumblebee has spread rapidly, and I saw one for the first time in my own garden in Sheffield in 2010.

Tree bumblebees were always present just across the English Channel, so their eventual colonisation of Britain is scarcely a surprise, although the fact that they took so long to arrive is. A much more remarkable colonist is the collared dove, a bird occurring originally across Asia from Turkey to China. Collared doves began to spread west in the nineteenth century, reaching the Balkans around 1900, Germany in 1945 and Britain by 1953 (where they were recorded in Norfolk as a breeding bird in 1955). Since then their spread has been remarkable, taking only two years to reach Scotland and a further two to reach Ireland. During the last quarter of the twentieth century numbers in Britain increased fivefold, but have been relatively stable since then.

MIGRATIONS, OCEAN DISPERSAL AND ISLANDS

There's no difficulty in explaining how tree bumblebees and collared doves reached Britain – both can fly and the English Channel is only a narrow waterway. Equally it's easy to see how horses and camels travelled from Kansas to Africa: they simply walked. On the other hand, species of all kinds are capable of migrations that are not always easily explained, and some of these take place quite routinely. Aphids, once they get high enough to be carried by the faster winds that occur further from the ground, can travel long distances. Hop aphids frequently turn up more than 100 km from the centres of hop production in southern England. Black bean aphids overwinter in Britain on the shrub spindle, which has a strongly southern distribution. Nevertheless, from this southern base they go on to colonise the entire country every spring. Plants can disperse long distances if they have small, light seeds; alternatively, they can hitch-hike on animals. Large herds of migrating herbivores, following the fresh plant growth arising from changing temperature or rainfall, were once widespread and still occur in some parts of the world. Transhumance, a domesticated analogue of the same process, also still occurs, albeit less commonly than in past centuries. In Spain sheep flocks are still shepherded along the traditional *cañadas* from the Cantabrian mountains

Modern transhumance – sheep on the move in the Drôme, France, 2011.

to Extremadura via Madrid every autumn. Even though the journey takes 28 days, Spanish ecologists Pablo Manzano and Juan Malo have shown that large numbers of seeds are easily transported the whole 400 km distance.

The existence of any life at all on remote oceanic islands indicates just how effective long-distance dispersal can be. *Metrosideros*, a genus of trees including the pohutukawa or New Zealand Christmas tree, also occurs on several Pacific islands. This genus originated in New Zealand, jumped over 5,000 km to the remote Marquesas Islands, and then another 3,000 km to Hawaii. The trees in all three places are genetically almost identical, so both jumps must have occurred some time in the last one to two million years. *Metrosideros* has tiny wind-dispersed seeds, so perhaps these remarkable journeys are not so surprising.

Similar journeys by animals occur, too, both through animals capable of long-distance travel, and by more accidental rafting on floating vegetation. Such rafts are not uncommon; one recent survey reckoned there are about 70 million kelp rafts afloat in the Southern Ocean at any one time. On Christmas Island, 5,000 km from North America, so much American redwood, fir and walnut is washed ashore that it's used as firewood. A Stone Age human culture subsisted on Pacific coral atolls, which lack any rocks of their own, solely on the basis of stones stuck in the roots of drifted trees. Even if few such rafts and logs have passengers, and if even fewer ever arrive anywhere, it's easy to see how significant numbers of colonisations might occur over geological time. The principal vertebrate raft-passengers are reptiles, since their waterproof skins and low metabolic rates (and hence low food requirements) fit them well for long sea voyages with little or nothing to eat. Both giant tortoises and iguanas on Galápagos have relatives in South America and are assumed to have rafted to the islands, a distance of some 960 km. Some even longer raft journeys are thought to have occurred; for example, Cuban geckos are thought to have rafted from the Mediterranean. One conspicuous feature of such long-distance dispersal is how unpredictable it is. Most of the animals on Galápagos may have come from the nearest source in South America, but Galápagos penguins came from the Antarctic, aided by the cold Humboldt Current, and the archipelago's sea lions came from California, apparently quite recently. Nor are long-distance travellers always the most intrepid species around. Amphisbaenians, or worm lizards, are a group of limbless, subterranean reptiles that rarely leave their burrows. Given their occurrence on both sides of the Atlantic, the natural assumption is that they evolved before the break-up of Gondwana and so didn't need to disperse anywhere. But recent molecular analysis shows that the lizards on opposite sides of the Atlantic are far too similar to have been separated for so long. They must have crossed the widening Atlantic, probably around 40 million years ago, a distance of around 5,000 km. The only available solution is rafting, which is surprising enough, except that it seems to have happened twice – once from tropical Africa to South America, and once from the Mediterranean to Cuba (maybe, in the latter case, on the same raft as the geckos?).

Some plants make similar journeys, or at least their seeds do, and they don't even need rafts. Salt water is quite quickly fatal to many seeds, but some are regular ocean travellers. Every postcard you ever received from friends on some tropical beach features an overhanging coconut palm, and with good reason: coconuts are well adapted for ocean dispersal.

Darwin was fascinated by oceanic dispersal of seeds, and tested the ability of many seeds to float and survive in salt water, finally concluding that 14 per cent of the seeds 'of any country might be floated by sea currents during 28 days across

924 miles of sea, to another country, and when stranded, if blown to a favourable spot ... would germinate.'

The seeds that survive best are those with waterproof seed coats, and such seeds frequently come ashore on British beaches, having been borne on the Gulf Stream from the Caribbean, the voyage taking a little over a year. Of course, those particular seeds do not find a suitable climate at their destination, but they serve to illustrate the principle. Over shorter distances, sea-borne seeds arrive in even larger numbers; many of the first plants to establish on the new Icelandic volcanic island of Surtsey, and on Krakatoa after the famous eruption, were from seeds transported by sea.

DISPERSAL BY HUMANS

Modern humans appeared about 200,000 years ago, and from the very start must have dispersed other species, if only by transporting viable seeds in their guts, like other herbivorous or omnivorous animals. They were also, of course, accompanied by their own collection of diseases and parasites – including the head lice we acquired from chimpanzees and the pubic lice that hopped across from gorillas. At first, seeds that attach themselves to animal coats must have found relatively hairless humans rather unrewarding dispersers, but all that changed when we first started wearing clothes, which may have been as long as 170,000 years ago. Nowadays, for plants with the right kinds of seeds, socks are a major dispersal pathway.

A step change in human dispersal occurred with the domestication of animals and crops, and the deliberate movement of these over large distances. This happened at different times in different places, but farming arrived in Britain about 6,500 years ago, partly through the physical migration of farmers from further south and east, and partly through the adoption of the new technology by existing residents. Sheep, goats, wheat and barley were imported from Europe, and inevitably these relatively few deliberate introductions were accompanied by many more accidental ones. Because seeds and pollen are often well preserved in archaeological deposits, our knowledge of plant introductions to Britain is particularly good, although of course still far from complete.

Before the clearing of woodland for pasture and crops, Britain was essentially completely wooded for several millennia at the start of the present interglacial (although there are those who would argue with that view), so it's not surprising that a very high proportion of Britain's arable weeds and other plants of unshaded habitats are known (or suspected) to be intro-duced, all probably accidentally by early farmers. Many very familiar plants first appeared in Britain between the Neolithic and the Iron Age, including (among many others) corn cockle, wild oat, burdock, mugwort, white campion, shepherd's purse, cornflower, corn marigold, hemlock, treacle mustard, henbane, common mallow, mayweed, annual nettle and four species of poppy, including the familiar cornfield weed. But not all plant introductions were accidental: two presumably deliberate Iron Age introductions are the dye plants woad (*Isatis tinctoria*) and weld (*Reseda luteola*).

Agriculture provided such a perfect habitat for plants of open, unshaded habitats that some plants may well have evolved specifically in response to this once-in-a-lifetime opportunity. The common red corn poppy, *Papaver rhoeas*, for example, appears not to have a 'natural' habitat at all; throughout its range it exists only in arable fields and similar man-made habitats. So successful were poppies and their other weedy contemporaries that their occupation of northern Europe was only the first step in a global journey. As farming practices that originated in southern Europe and the eastern Mediterranean spread around the world, so did the weeds that evolved alongside those practices. At least outside the tropics, European weeds rule the world.

Considering their relatively brief tenure, the Romans introduced a long list of plants to Britain. This includes another long list of weeds (basically most of the suitable candidates not present already), plus millet, lentil, fig, olive, medlar, pear, almond, peach, garlic, onion, shallot, leek, cabbage, pea, cucumber, lettuce, turnip, radish, asparagus, rosemary, thyme, bay, basil, walnut, sweet chestnut, 'real' apples (as opposed to the native crab apple), grapevine and mulberry. It's a testament to its subsequent importance as a dietary staple that the leek, a Roman introduction, was later adopted as the national emblem of Wales. If we add guinea fowl, chickens, rabbits, brown hare and pheasants to the long list of plants, it's clear that food as we know it in Britain really only began with the Romans.

Even so, it seems Roman taste wouldn't be entirely familiar to a modern palate. The Romans seemed particularly fond of umbellifers, or members of the carrot family (but not the carrot itself; the familiar cultivated carrot – as opposed to wild carrot – is a much later introduction). Among that family, the Roman introductions parsley, celery, fennel and dill are reasonable enough, but few people today would consider the Roman pot-herbs alexanders and ground elder to be particularly appetising, or even edible at all. In medieval times plant introductions continued, including caraway, chicory, wallflower, soapwort, horseradish and yet more weeds. The impact of all these introductions on the 'wild' British flora has been very variable; many of the food plants survive only in cultivation, but many of the weeds are still common and, as any gardener will testify, ground elder is depressingly abundant, while sweet chestnut is now so thoroughly naturalised that it was long assumed to be a native tree.

What did the Romans do for Britain? Apart from introducing chickens, rabbits, apples, pears, onions ... and Welsh leeks (and the thyme, too).

After about 1500, the discovery by Europeans of the New World and a general increase in voyages of discovery, and ultimately colonisation and trade, greatly increased the distances over which humans were able to move animals and plants. As a result, German and French botanists proposed over a century ago that introduced plants known (or presumed) to have arrived in northern Europe before 1500 be distinguished from those that arrived later. The former were dubbed archaeophytes and the latter neophytes, terms that are still widely used today. Unsurprisingly, most of the archaeophytes, as accomplished fellow travellers of agriculture and human settlement, quickly also found their way to North America, Australia and New Zealand. So quickly, in fact, that many early human colonists must have wondered whether these familiar plants had actually been there all along, and it was only later naturalist-explorers who realised they had been introduced.

Peter Kalm, dispatched to America by Carl Linnaeus in 1747 in search of economically useful American plants, found (among others) hemp, privet, ivy, wild parsnip, dandelions, purslane, white clover and greater plantain. The latter, a consummate colonist of disturbed, trampled ground, was such a consistent feature of European settlement in North America that it was known by Native Americans as 'white man's foot'. Kalm was a botanist, but he also couldn't help noticing some other European imports that he wished had stayed at home: bedbugs, houseflies and cockroaches. On a happier note, he also noted honeybees, writing that 'The Indians ... generally declare, that their fathers had never seen any bees either in the woods or any where else before the Europeans had been several years settled here [the Philadelphia area]'. In recognition of their origin, Native Americans named honeybees 'English flies'.

The plants and insects noted by Kalm were only the start. As human numbers have grown and trade and travel increased, ever larger numbers of species have been moved between countries and continents. Today it's almost impossible to guess how many, although that uncertainty is much greater in some groups than others. Mammals are generally big enough to be easily noticed, and mostly they have been introduced deliberately (although there are some notable accidental introductions, such as the house mouse, a native of the grasslands of central Asia but now found wherever there are people). Thus we know, for example, that 81 mammal species have been introduced into the wild in the USA. Or at least we think we do – that number apparently includes bison (an iconic US 'native' if there ever was one), horses, camels and chimpanzees. We have indeed returned horses to North America, where they started out, but although camels were used as pack animals by the US Army in the 1850s and a few later escaped into the wild, none seem to have made it into the twentieth century, and you would have to try pretty hard to find a wild chimpanzee in the USA. In any case, lists of anything introduced into the USA always have to be approached with great caution, since such lists usually include Hawaii, which is far more invaded than the continental US. A remarkably similar number of mammal species (88) have been introduced into Europe, but currently only 59 survive in the wild. Most of these are from North America or Asia, but a few represent introductions from one part of Europe to another.

At the opposite extreme, invertebrates are far more likely to escape attention, and most are introduced accidentally. So the best we can say, for example, is that by 2010, 1,590 alien invertebrates were identified as established in Europe, nearly all of them insects, with contaminants of ornamental plants providing the single most important route. We know most of these insects eat plants, and that most come from Asia (with North America a close second). On the other hand, we haven't a clue what proportion of introductions those successful species represent, though all the evidence suggests it must be very small; almost all insect introductions fail. One indication of that failure is that even the successes generally don't make much headway; most of the alien insects established in Europe have remained confined to the country where they were first introduced, and in those countries the overwhelming majority (more than three-quarters) are found only in man-made habitats, such as agricultural fields, parks, gardens and buildings. A significant number are still stuck in greenhouses, presumably because they came from warmer climates and can't survive outside.

Plants tend to occupy an intermediate position between mammals (about which we know a lot) and insects (about which we know very little). For example, a substantial fraction of the world's temperate flora must be growing in Britain somewhere, if only in one or two botanical gardens, but how many have escaped into the wild? Here the problem is not so much counting the number of introductions, which is relatively easy, but the subtle gradations of 'escaped into the wild'. A widely accepted rule of thumb in Britain is that 10 per cent of all plants imported into the country go on to escape to some extent. Of these, around 90 per cent remain as 'casuals', that is they survive in the wild only as long as they continue to escape from cultivation; if this source of colonists is cut off, they die out. The other 10 per cent establish genuinely self-sustaining populations in the wild, and about 10 per cent of these go on to be regarded (at least in some quarters) as pests. In terms of absolute numbers, probably about 12,500 species of plants have been introduced into Britain, but only about 200 count as fully established, in the sense of looking more or less like natives. The number of alien plants troublesome enough to be described as pests depends on your level of tolerance, but may be as few as 11 or as many as 39.

WHAT A LONG, STRANGE TRIP IT'S BEEN

The world, and the history of its inhabitants, are both much weirder than most people realise. Three hundred million years ago, any discussion of species dispersal across oceanic barriers would have seemed pointless, since all the world's land was crammed together into a single continent, Pangaea. As a result, the diverse fauna of amphibians and reptiles that dominated the world at the time was more or less the same everywhere; nowhere was more different from anywhere else than New York State is from Oregon today. The subsequent history of the Earth is essentially of the coming apart of that original continent, which took a very long time; only about

50 million years ago would a view of the Earth from space have started to look vaguely familiar. Nor has any of this stopped, or even slowed: the landing place of the *Mayflower* Pilgrims, Plymouth Rock, is fifteen metres further west than it was in 1620.

One consequence of all this is that some species have ended up a long way from their ancestors, or from their current close relatives, merely by staying in the same place, as the Earth has moved beneath them. But mostly species have moved, often because they had to: when the environment changes, migration is easier and quicker than evolution. Just before the start of the present cold period, hardly last week in geological terms, lush forests extended to Alaska and crocodiles lived in the Thames. Since then, the ice has obliterated the life of the north 16 times, covering up to a quarter of the Earth's land. It's easy to assume that only colder areas were affected, but the climatic impacts of glaciations were global. At the last glacial maximum, rainfall in the Amazon basin was lower by nearly a half and the Amazon rainforest was reduced to isolated refugia, separated by large tracts of grassland, savannah and, in the driest parts, semi-desert. Vast deep lakes, up to the size of Lake Michigan, repeatedly came and went in America's arid Great Basin region. Each time life has recovered, but never the same species twice, and certainly never the same genes. At any spatial or temporal scale you care to examine, history never repeats itself. No one knows how many different snails used to live on Krakatoa, but five species were collected before the famous volcanic eruption. The modern island has 19 different snails, but none of that original five is among them.

In short, the Earth is a dynamic, dangerous and unpredictable place. Any given spot is generally, and usually sooner rather than later, subject to some extreme geological or climatic catastrophe. Species that do not or cannot move go extinct, or at least their numbers are greatly depleted. In the struggle for existence, natural selection favours those species that are able to move in response to a changing environment, or which already did so and have thus come to occupy a large range. Nowhere is stable, nowhere is isolated, and they never have been. Darwin knew this, as is clear from *The Origin of Species*: 'Considering that the several above means of transport, and that several other means, which without doubt remain to be discovered, have been in action year after year, for centuries and tens of thousands of years, it would I think be a marvellous fact if many plants had not thus become widely transported.' Darwin was talking about plants, because he had his own experiments with seeds to draw on, but he clearly had no doubt that much the same applied to animals.

We have short memories, and we are fortunate to live in a brief moment of relative climatic calm (although it seems an unimaginably long time to us), so we tend not to notice any of this. We are happy to accept that wherever species happen to be right now is where they belong, and where they ought to be. Against this background, of a fairly fundamental disconnect between the world as it is and the world as we sometimes imagine it to be, how has mankind's attitude to 'nativeness' and 'belonging' developed, where are we now, and where are we headed? Those are the questions considered in the next chapter.

CHAPTER TWO

A SHORT HISTORY OF NATIVENESS

The idea that some plants and animals are found in one place, and others somewhere else, developed along with the ability to recognise and name different species. Clearly, one could not say that species X differed in its distribution from species Y until those species could be reliably and consistently named and identified. So it's perhaps no great surprise, as we saw in the previous chapter, that one of the first scientific observers of introduced species was Peter Kalm, a protégé of Carl Linnaeus. Linnaeus, of course, was responsible for introducing the standard Latin binomial names for plants and animals that we still use today. However, neither Linnaeus nor Kalm would have recognised the concept of *native* as it is understood today; until the end of the eighteenth century 'native' simply meant 'wild', i.e. untamed, uncultivated or undomesticated.

WHAT IS NATIVE?

The modern division of species into native and alien first appears in the writings of Hewett Cottrell (H. C.) Watson in the mid-nineteenth century. Watson, a keen amateur botanist, was aware that previously unrecorded species were appearing in Britain, and he thought some sort of classification system was needed to keep track of such species. He was the first to define 'native' in the modern sense: 'apparently an aboriginal British species; there being little or no reason for supposing it to have been introduced by human agency'. But although his definition was modern, his attitude to alien species was not, in that his problem with alien species was essentially scientific. That is, he thought native plants would show reliable correlations with soil and climate, but that trying to do the same with recent immigrants would only cloud the picture. Aliens weren't inferior; they were just less interesting.

Watson also had a curiously nineteenth-century motivation: he hoped that the establishment of a definitive 'native' British flora would prevent competition among 'vainglorious' botanists to discover new native species, even to the extent of planting these 'discoveries' themselves. Crucially, he also recognised that absolute certainty about native or alien status would often be lacking, and that some long-established aliens were *de facto* natives: 'Species originally introduced by human agency now exist in a wild state; some ... continued by unintentional sowings ... while several keep their acquired hold of the soil unaided, and often despite our efforts to dispossess them. Both these classes certainly now constitute a part of the British flora, with just as much claim as the descendants of Saxons or Normans have to be considered a part of the British nation.'

Much has been written on the subject in the last 150 years, but Watson's definition of 'native' still stands. There is much more to say about the *value* of nativeness, but it's worth pausing here to

The originator of 'native' and 'alien' species, Hewett Cottrell Watson.

ponder the usefulness of the definition itself. That 'introduced by human agency' seems simple enough, but is it? The first question is one of timing. Does human introduction always render a species alien, or (in Britain, for example) is it only post-Neolithic, post-Roman or maybe even post-medieval introductions that are 'really' alien? Watson considered 'native' and 'alien' to be absolute categories, not susceptible to subdivision, but, as we've seen, European botanists were already distinguishing between pre-1500 (archaeophyte) and post-1500 (neophyte) introductions. The former are often accorded 'honorary native' status, and government bodies may even devote considerable time and money to conserving them.

Uncertainty about the importance of timing prompts a further question: what does extinction, followed by human reintroduction, do to a species' native credentials? The original native populations of white-tailed eagle, goshawk and capercaillie are extinct in Britain and have been successfully reintroduced by man, so the present birds are human introductions, and are genetically distinct from the extinct natives. Are they, nevertheless, still *bona fide* natives? Of course, they're quite recent extinctions, so what about reindeer, extinct in Scotland for 8,000 years but now reintroduced? And if they're all native, how far *do* we have to go back before a reintroduction is no longer native? Even the mandarin duck, now confined to eastern Asia, is a British native if we go back only a little further, though here at least the law is clear: the mandarin duck appears in Schedule 9 of the Wildlife and Countryside Act 1981, making it illegal deliberately to release or negligently to permit this former British native to escape into the wild in Britain.

Even asking these questions begs a much larger one: what is so special about human agency? Species introduced by man are supposedly different from those dispersed by any other agency because man became, at some point, no longer part of 'nature', but something quite different. But plenty of people have held the view that this has never happened at all, and that humans continue to be bound by the same rules as the rest of the living world. Here, for instance, is Charles Lyell, pioneering geologist and close and influential friend of Charles Darwin, in his most famous book, *Principles of Geology*:

> We have only to reflect, that in thus obtaining possession
> of the Earth by conquest, and defending our acquisitions
> by force, we exercise no exclusive prerogative. Every species
> which has spread itself from a small point over a wide
> area, must, in like manner, have marked its progress by the
> diminution, or the entire extirpation, of some other, and must
> maintain its ground by a successful struggle against the
> encroachments of other plants and animals.

And even more explicitly:

> ... *we may regard the involuntary agency of man as strictly analogous to that of the inferior animals. Like them we unconsciously contribute to extend or limit the geographical range and numbers of certain species, in obedience to general rules in the economy of nature, which are for the most part beyond our control.*

In other words, in a rare expression of humility in Victorian Britain, we are not as unique (or as clever) as we think we are.

But even if we agree that man is indeed now 'outside nature' (a dangerous idea if ever there was one), and we can agree the date on which we entered upon this transcendent state, there remains the vexed question of how much 'human agency' it takes to make a species alien. In the previous chapter I mentioned the recent colonisation of Britain by the tree bumblebee and collared dove. They may seem to have little in common, beyond the ability to fly, but both blur the distinction between natural and human-assisted introduction. As far as we know both species reached Britain on their own, without direct human help, from areas where they were already native, so technically that now makes them both native British species. But in neither case can indirect human influence be ruled out. Both species seem to like people, and especially gardens. Spilled farmyard grain, animal feed and food put out for birds by gardeners are strongly implicated in the spread of collared doves, and they are now one of the top ten most common British garden birds. Similarly, tree bumblebees are abundant in gardens, with many British gardeners waking up recently to find their bird nest boxes colonised by this bee, which naturally nests in holes in trees.

Can it be coincidence that these species, both of which thrive around humans, waited 10,000 years after the end of the last glaciation to colonise Britain? In fact, tree bumblebees

Human agency at work? An 'alien' collared dove feeding at a bird table in an English garden.

and collared doves are just two examples of a wider problem, which is the wholesale transformation of the landscape by man, making it far less favourable for some species and much more favourable for others. If the former happen to be native and the latter alien, how much 'human agency' is involved in the replacement of the one by the other?

Not only is the world now physically and chemically transformed by man, whole ecosystems are on the move in response to climate change, itself now widely accepted to be largely of human origin. So to argue in the twenty-first century that any contraction, expansion or shift in the range of any species is independent of human agency is to make an assertion that, almost by definition, can rarely be literally true.

WAR AND PEACE

H. C. Watson may have given us our modern distinction between aliens and natives, but he did not have any fixed opinions on their relative merits. Over the following century, as more and more species were moved, more and more naturalists and travellers noticed and wrote about the changes that followed. At the same time, mankind began to have ever-greater effects on the environment as the human population increased and demand grew for energy, minerals, living space and, above all, food and water. If native biodiversity declined, while at the same time alien species thrived, some observers were inclined to blame the latter for the former, and in some cases to issue blanket condemnations of introduced species. Others took a more nuanced view, recognising that introduction of alien species rarely took place in a vacuum, and that there were usually other things going on. For example, in his monumental *The Naturalisation of Animals & Plants in New Zealand*, published in 1922, George M. Thomson wrote:

It must not be supposed that it is the introduced animals alone which have produced [the retreat of the natives], even though rats, cats, rabbits, stoats and weasels, as well perhaps as some kinds of introduced birds, have penetrated beyond the settled districts. It is largely the direct disturbance of their haunts and breeding places, and the interference with their food supply, which has caused this destruction and diminution of the native fauna ... [m]any insects which were common in the bush fifty years ago must have been displaced and largely disappeared. I cannot appeal to figures, but the surface burning of open land which prevailed, especially in the South Island, and the wanton destruction and burning of forest which has marked so much of the North Island clearing, must have destroyed an astonishing amount of native insect life, and made room for introduced forms. The clearing of the surface for cultivation and grazing, the draining of swamps, and the sowing down of wide areas in European pasture plants, have all

contributed to this wholesale destruction and displacement
of indigenous species.

Thomson recognised, as did many others, that alien species were often very good at filling the gap left when native species (and entire native ecosystems) were destroyed, but that the aliens were rarely the underlying cause. In particular, he noted that introduced weeds depended almost entirely on human settlement and agriculture:

> *The opinion of all botanists in New Zealand to-day is that*
> *when the direct, or – to a large extent – the indirect influence*
> *of man is eliminated, the native vegetation can always*
> *hold its own against the introduced. Those plants which*
> *have thriven abnormally in this new country, and have*
> *impressed visitors by their abundance, are found in settled*
> *and cultivated districts, and belong chiefly to what are known*
> *as weeds of cultivation, that is, plants which have become*
> *adapted to conditions caused by the direct and indirect*
> *actions of human beings, and which only thrive where those*
> *conditions are maintained.*

Both views – that aliens are themselves a major cause of environmental harm, or that they are merely symptoms of other changes – coexisted throughout the century after Watson. Ecology textbooks reflected both viewpoints, but rarely had much to say on either side.

All this changed in 1958 with the publication of Charles Elton's book *The Ecology of Invasions by Animals and Plants.* Today's invasion biologists, if questioned, generally claim Elton's book as their inspiration, and it has been described as signalling 'the beginning of the field of invasion biology', and as 'a bible for practitioners of a burgeoning new science'. But in many ways it is an odd book. It isn't a scientific book in the usually accepted sense, nor is it a textbook. It is in fact a popular polemic, based to a large extent on a short series of radio talks that Elton made for the BBC. But what is not in doubt is that it sits squarely in the tradition of blaming introduced species for practically any environmental ill you care to mention, views that Elton held long before 1958. For example in 1944 Elton had written:

> These are major engagements in a violent struggle against the spread of undesirable plants and animals that is affecting every country. As I have pointed out elsewhere we are witnessing not only the immediate dislocations caused by the introduction of various species into countries new to them, but a vast historical event – a zoological catastrophe, which is the beginning of the breaking down of Wallace's zoogeographic realms and innumerable island isolations by the activities of man, a process which will eventually reduce the rich continental faunas to a zoned world fauna consisting of the toughest species.

Elton had lived through two world wars, so 'invasion' was a more loaded idea for him than it is for you or me. Indeed, in the Second World War he worked on controlling rabbits, rats and mice – 'alien invaders' that were eating stored food and farmers' crops, and were thus practically in league with the Nazis. The result was a remarkably and unashamedly militaristic book, as in this not atypical sentence:

> I have described some of the successful invaders establishing themselves in a new land or sea, as a war correspondent might write a series of dispatches recounting the quiet infiltration of commando forces, the surprise attacks, the successive waves of later reinforcements after the first spearhead fails to get a foothold, attack and counter attack, and the eventual expansion and occupation of territory from which they are unlikely to be ousted again.

Another curious feature of Elton's book, or at least of the reaction to it, is that the 'burgeoning new science' to which it gave rise took a remarkably long time to do much real burgeoning. Interest in 'invasions' by alien species, as measured by the number of academic publications on the subject, trundled along after 1958 at much the same very low level as before, with the subject being scarcely mentioned for the next 30 years. But academic interest did slowly begin to grow, and it accelerated in the early 1990s, fuelled by the advent, in 1999, of the specialist journal *Biological Invasions*. Public and official government concern grew alongside this academic interest, with (for example) a US Presidential Executive Order in 1999 calling for the formation of a National Invasive Species Council to render the federal response to introduced species more effective. Today most governments take very seriously indeed their responsibility to protect us from introduced species, spending large sums on research and colossal sums on the Sisyphean task of control. I use that adjective deliberately, since, like the labours of Sisyphus himself, most of this money, as we will see, is wasted.

THE VALUE OF NATIVENESS

It's clear that Elton's legacy was not a new science, but a set of old attitudes, founded on principles left over from before Wegener and plate tectonics, and even from before Darwin. Principles founded on a fixed flora and fauna inhabiting a static and unchanging world. In many ways Elton seems to have longed for the lost innocence and certainty of childhood, writing early on in *The Ecology of Invasions*: 'When one was a child, [the distinct faunas of different realms were] very simply summed up in books about animals. The tiger lives in India. The wallaby lives in Australia. The hippopotamus lives in Africa.' It

Charles Elton on a wartime survey trip, with a bag of mousetraps.

was apparently neither here nor there that the hippopotamus had (relatively recently) lived in what is now central London. Or that tigers had first arrived in India even more recently, or that they had very nearly made it to Alaska, and thus come within a whisker of being bona fide US citizens.

Elton believed firmly that species belong to wherever they happen to be right now, irrespective of length of tenure or of where they had evolved or migrated from. More than that, he believed that belonging confers rights of occupancy, that these rights extend indefinitely into the future, and that natives are morally superior to aliens. And these are views all too often shared by Elton's many modern admirers and disciples.

As New Zealand invasion biologist James C. Russell wrote in 2012: 'Our ethical duty to non-native species ... differs from our duty to native species.' This conferring of a moral superiority upon natives, and the ethical obligations that arise from it, do not appear to derive from any obvious source, in Elton's writings or anywhere else; they are simply values, like a belief in free speech or democracy, that are not seen to require any justification.

Interesting as the origins of giving rights to natives that are denied to aliens are, perhaps even more interesting is why this idea persists today. Two motives occur to me, one rather more noble than the other.

THE CONSERVATION IMPERATIVE

Nature conservation seems such a thoroughly laudable and well-intentioned activity that it appears almost churlish to impugn its motives. But questions that must occasionally occur to all conservationists are: 'what should we conserve, and why?'. These are not simple questions, and in so far as the answers exist, they may be radically different in different places.

In the USA, the benchmark is pre-Columbus, which is seen (Native Americans notwithstanding) as essentially unmodified by man. The major conservation instrument is the network of National Parks, within which the wildlife, as far as possible, is left to get along by itself. In the UK, there is no such easy recourse to a year zero, since almost all the landscape has been managed for millennia, which also means that leaving things alone is not an option. Broadly, the chosen objective in the UK (and particularly in England) is the agricultural landscape that would have been familiar to Jane Austen. But in both countries (and elsewhere), one powerful criterion of conservation-worthiness is the presence of native species, or more specifically the absence of alien ones.

Conservation priorities in the UK can be traced back to the 1977 *Nature Conservation Review*, written by Derek Ratcliffe, Chief Scientist for the then Nature Conservancy Council. Ratcliffe established a list of criteria for valuable wildlife sites, one of the most important of which was 'naturalness': essentially a measure of the amount of modification by man – where less modified, more 'natural' areas have a higher nature conservation value. Naturalness is at the same time one of the most important, and also one of the most vague, of Ratcliffe's criteria – hardly surprising in a country where practically nowhere can lay claim to being even slightly natural.

Since 'naturalness' itself could hardly be defined, let alone measured, the search was on for surrogates that indicate degrees of naturalness. Step forward alien species, by definition the result of human interference, and thus useful indicators of 'unnaturalness'. As Ratcliffe put it, there was 'a general feeling that plants deliberately introduced . . . into a country, or even a new locality within a country, have lower status than those which arrived of their own accord, and that they have no particular claims in regard to conservation'. It's a short step from the aliens themselves having no claim to be conserved to their presence being an indicator of a site that has no such claim.

Nor, until recently, were alien species admitted to the British *Red Data Books* that list threatened species. That prohibition has now been (somewhat) relaxed, but aliens admitted to the conservation tent must still be judged 'native enough', which in practice means (some) archaeophytes are acceptable, but neophytes are not. Thus a refusal to consider aliens as worthy of conservation, intended to bring a bit of rigour to the otherwise subjective topic of what should be conserved, has in practice the status of a rule that was made to be broken. And if an attractive, well-behaved neophyte manages to become part of an otherwise valued semi-natural habitat (as in an example I describe in Chapter 5), then the rules have to be broken a bit more. In short, aliens have no conservation value … except when we decide they have.

FOLLOW THE MONEY

If you wanted to be rich and famous, then you wouldn't become a scientist in the first place. But if the limit of your ambition is a busy research lab, full of postdocs and students carrying out well-funded research projects, and plenty of invitations to speak at global conferences (all reasonable enough aspirations, surely), then you have to work on something that society thinks is important, or at least worth funding. Which boils down to attracting the attention of politicians. And what politicians, who naturally enough want to be re-elected, are inclined to think is important is whatever their constituents think is important. Which ultimately is what newspapers, television and online news sources tell them is important.

So what is that? Well, here's David Shukman, the BBC's environment correspondent: 'Experts in any field can get caught in a vortex of anxiety. And the news media are receptive, with an appetite for scares. Missile gaps, dodgy chemicals, creepily-modified food, monster asteroids – all can be big stories because editors think rightly that they'll fascinate readers and viewers.' Shukman is here musing on how to respond as a journalist to climate forecasts. His list of scares doesn't include invasive alien species, but easily could have done. And here are the thoughts of nine American invasion biologists:

In the United States, hardly a day passes that media attention is not somehow focused on an invasive species issue, variously involving legislatures, mayors, governors, state and federal agencies, the courts, concerned citizens, and advocacy groups on all sides. These concerns range across the real and growing problems of economic damage, ecosystem degradation, and competition with rare or desirable native species and the real or perceived threats of such impacts. A growing technical literature – papers, books, online resources, and the associated nontechnical, educational resources – has helped fuel both scholarly interest and public concern.

And there, in a nutshell, lies the problem. Alien species seem practically designed to excite public concern. Almost by definition they are most abundant, and most visible, in the most highly human-modified habitats, such as towns and cities. Personal encounters with aliens are routine, so everyone has an opinion, and it's often 'obvious' that aliens are actively supplanting natives, even if that isn't what's happening at all. It's equally 'obvious' that something must be done, even if it's not clear what that should be, and even if ill-judged intervention might only make things worse. So, there's a journalist on the phone, and he wants an answer now: 'Will the Asian giant hornet invade Britain and wipe out all our honeybees?' Privately, you think both of those propositions are pretty unlikely, but if it's not a threat why would anyone pay you to go to the south of France and study the critter? In any case, by the time you've explained that most introductions fail completely, that hardly any of those that succeed go on to cause any kind of problem, and that among those that do it's rarely the ones anyone worried about beforehand, you know you'll be talking to yourself. On balance, it's much easier to admit it's not impossible, thus contributing to tomorrow's news headline.

Note that no one needs to be actively dishonest in order to stoke public paranoia; all that's necessary is to admit there are a handful of genuine invasive alien horror stories, that it's conceivable that your favourite study species could one day be among them, and generally to be unwilling to give your journalist interlocutor the short lecture course in basic biology and how to distinguish correlation from causation that he so clearly needs but is unlikely to absorb. If, as a result, you end up surfing a wave of popular concern, rather than becalmed with all the dead seaweed and other dreary flotsam, then that's hardly your fault, is it?

THE REST OF THIS BOOK

You are by now (I hope) at least partly persuaded that deciding where species 'belong' is not quite as simple as is popularly suggested. Despite that small problem, and despite 'nativeness' being an elusive and certainly an immeasurable quality, we live in a world where it has been elevated to a primary concern, if not *the* primary concern, of conservationists. A world where we are engaged in a life-and-death struggle against a rising tide of invasive species that threatens to engulf us all. And if you doubt that for a moment, here (proving not only that there is no limit to hyperbole in the service of a bad argument, but that hyperbole is often the sign of a bad argument) are some quotes from a recent scientific paper on the invasions threat:

The global damage inflicted by invasions amounts to $1.4 trillion per year, which constitutes 5 per cent of the global economy and is nearly an order of magnitude higher than the annual global cost of natural disasters.

Invasive species are also a major cause of the declines and extinctions of native species.

Invasions also resemble catastrophic accidents in high-tech industries, e.g., nuclear power ... Preventative management of invasions like that of natural disasters requires international coordination of early-warning systems, immediate access to critical information, specialized training of personnel, and rapid-response strategies.

None of these statements can be in any way justified. Yet, bearing in mind, as we will see later, that eradication of alien species is extremely problematic except on a few islands, and even then under only the most favourable circumstances, what would taking such advice at face value involve? We would face a world locked into an interminable, unwinnable and ruinously expensive war on introduced species. The difficulty of ensuring there are never any unwanted passengers in (or on) planes and ships, or in their cargoes of fruit, grain, plants or timber, would make international trade hugely expensive and difficult; indeed, it might have to be banned altogether. Airport security for ordinary passengers would almost certainly have to be significantly tougher.

I have provided so far a very brief sketch of how we arrived at this absurd situation. In the rest of this book I look at how the intellectual quagmire that is invasion biology is maintained and policed by its adherents, and take a deeper look at nativeness. I reveal a reliance on old, unreliable and often totally discredited data, which continues to be believed as long as it supports the dogma that 'the only good alien is a dead alien'. I show that we understand very little of the basic ecology of invasions, that the costs of alien species are routinely and grossly exaggerated, that fear of invaders is actually hampering conservation, and that many invasion biologists don't yet seem to have caught up with Charles Darwin. Finally I make a few modest suggestions about what we should actually be doing about alien species, or at least about the sort of evidence we should accept before we need to do anything.

I was going to say that along the way I debunk the myth that alien species constitute the second largest global threat to biodiversity (after habitat destruction). But on reflection I think that assertion has been debunked so often (yet is endlessly repeated) that it no longer deserves the status of a myth, and is best described merely as a straightforward lie. Not only are alien species not the second biggest threat to biodiversity, they aren't the third, or fourth, or even the fifth, largest cause.

Nor are invasive species one of the most important threats to the continued survival and well-being of mankind. The lives of large numbers of people are all far more likely to be ended or made miserable by war, disease, pollution, climate change and shortages of energy, minerals, food and – above all – fresh water. You will read a good deal in the next two chapters about tamarisk, one of the USA's most feared invaders, but before you do, bear in mind that none of this is actually about tamarisk. In fact, it's all about water – water and our deep-seated desire as a species to blame someone, or something, for our own short-sighted and inept stewardship of the environment.

CHAPTER THREE

FIRST SOME
BAD NEWS

I ended the last chapter by suggesting, among other things, that the threat from alien species is often much exaggerated. To illustrate this point I devote the next two chapters to an examination of two sides of the same coin, starting in this chapter with the popular view of four notorious alien invaders: the brown tree snake, zebra mussels, tamarisk and purple loosestrife. If you were to Google any of these four you would discover hundreds of thousands of hits, almost all describing their appallingly bad behaviour. Read a representative sample – and the pages that follow – and you will be seriously alarmed, and, I suspect, persuaded that the huge sums currently being spent on attempts to control them is money well spent; indeed, you might well think it's scarcely enough.

In the following chapter, I look again at the same four villains, but this time seeking out conflicting evidence. After reading both chapters you should then be in a position to make up your own mind about all four.

BROWN TREE SNAKE

Guam is one of the Mariana Islands, which lie about halfway between Japan and New Guinea. Its main claim to fame is the epic battle in 1944, during which it was recaptured from the occupying Japanese by US forces.

Guam is about 45 km long, mainly forested, but with some open savannah in the south. It has, or had, 22 native birds, plus several introduced species. During the 1960s, most of these birds began to decline, first in the south of the island, but later in the north too. The decline was particularly severe in forest species, and by 1983 all of Guam's forest birds were confined to a small strip of land in the extreme north. Ten species of forest birds are now extinct, with the remaining three very rare.

By the mid-1980s Guam was (and remains) a bird desert, a 'silent spring' of the sort Rachel Carson warned about in her famous book of the same name. But what was the cause? There was no shortage of theories. The cause of Carson's silent spring was pesticides, so maybe that was the case here? In the past, and particularly during the war, pesticides had been widely used on Guam to control mosquitoes and other insects, but analysis of the tissues of a variety of animals did not reveal unusually high levels of organochlorines or of anything else. Hunting seemed unlikely too; Guam's native birds had long been protected, and tight security prevented any hunting on military land, yet the decline was as bad there as everywhere else. An extensive study in the 1980s also ruled out disease as a factor.

Maybe Guam's native birds were losing out in competition from introduced birds? But the most likely culprit, the aggressive black drongo, did not seem to compete with native birds, and anyway is most common in the north, where the natives hung on longest. A final suggestion was loss of habitat, a major cause of animals and plant extinctions worldwide. But Guam is not unusual in this respect, and other islands that have fared worse, such as Saipan, have not seen anything like the same bird decline.

So that left just one other logical possibility – predation. Guam's birds had disappeared because something has eaten them. And because the decline had been so recent and so swift, an introduced 'something' was implicated. Cats, dogs, rats and monitor lizards have all been introduced to Guam, but feral cats and dogs are rare. Rats have certainly been involved in bird declines on other islands, including New Zealand and Hawaii. But rats are surprisingly rare on Guam – something that, as we shall see, is an interesting piece of evidence in the mystery. Finally, monitors have been on Guam for centuries, and birds and their eggs form only a tiny part of its diet. Not only that, all four predators have been widely introduced into other islands in the Marianas, none of which has experienced the same bird decline.

Suspicion thus fell on *Boiga irregularis*, the brown tree snake. Native to Australia and New Guinea, the snake arrived in Guam as a stowaway in military cargo in the late 1940s or early 1950s, and is the only bird predator unique to Guam in the Marianas. In fact if snakes had fingers, the brown tree snake's fingerprints would be all over this particular crime scene. Newspaper articles, field records and, crucially, local people all confirm that the snake started off in the south of the island in the early 1950s and spread north at about 2 km per year, reaching the far north in the early 1980s. Many local people are involved in farming and paid close attention to the sudden appearance of a previously unknown large snake (up to 2 m long, or occasionally more); Guam has only one native snake, which is small and rarely seen. The pattern of spread of the snake corresponded exactly to the decline in birds; soon after the snake arrived in a new area, the birds disappeared. A small offshore island, Cocos Island, makes a neat natural experiment – the snake is absent there, and the birds are fine. Confirmation of the role of the brown tree snake came from setting traps, baited with quail. This experiment took place in 1984, towards the end of the conquest of Guam by the snake, and revealed very high levels of snake predation right across the island, but lowest in the north, where the snake had only recently arrived. Quail in traps on Cocos Island were lucky – they all survived.

As well as birds, brown tree snakes eat small mammals and lizards, so what happened to them when the snake arrived? Comparisons of numbers of small mammals before and after the arrival of the snake reveal a big effect on these alternative prey, with declines of 90 per cent or more. Until the mammals were subject to proper scientific study, no one had noticed this decline, since the mammals are secretive and mostly nocturnal, so their decline was much less obvious than that of the birds. The only place high densities of mammals survive is in savannah, not a favourite habitat of the brown tree snake, which (as its name suggests) prefers forest.

When the brown tree snake first arrived in Guam, it encountered a paradise for a bird-eating snake, especially one that is strongly arboreal and largely nocturnal, and thus able to consume eggs, nestlings and roosting adults. Moreover, Guam's native birds, used only to a snake-free world, had few defences. Small, highly gregarious white-eyes roost together on a branch, shoulder to shoulder, and can be taken one at a time without the others taking flight, in a kind of ornithological kebab. Most other species soon followed, and the only birds to survive now are those that inhabit wetlands, savannah or urban areas, where snakes are fewer.

While the birds lasted, brown tree snakes achieved peak numbers of 100 per hectare, an almost unprecedented density. For snakes of their length, brown tree snakes are unusually

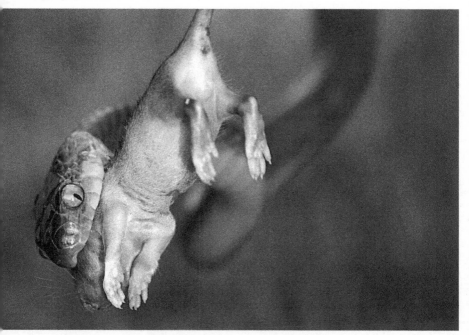

Not a pretty sight. A (captive) brown tree snake feeds on a rat.

slender, but quite suddenly, in the 1980s, people started to find snakes that were even skinnier than usual, with a lean and hungry look about them. The reason was, of course, that they had run out of birds. You might think this would be bad news for the snakes, and to some extent it was, with densities now well below the days when there was an easy meal on every branch. But brown tree snakes are adaptable animals, and they are also happy to eat lizards and small mammals, many of which are also introduced to Guam, including skinks, geckos, rats and house mice. Indeed young snakes subsist mostly on small lizards even when there are birds around. Thus any hope that the snake population would collapse when the birds ran out was destined to be disappointed. We also now know why Guam has fewer rats than you would expect.

The loss of almost all of Guam's birds is clearly a disaster, and the brown tree snake charge sheet has other items on it. The snakes are fond of pets and get through quite a few puppies and cage birds. They also render keeping poultry difficult if not impossible, so Guam has to import a lot of eggs, at considerable expense. Brown tree snakes are venomous, but they do not actually inject venom, so adults who are bitten can avoid a large dose simply by removing the snake. Infants are more at risk and occasionally end up in intensive care after being bitten, though there have been no fatalities. Snakes climb pylons and frequently short out power lines, causing power cuts. In fact somewhere on Guam experiences a snake-related power cut about every other day, and the fact that the snake in question is now very dead is little compensation for the serious inconvenience to both business and residential customers.

But apart from fixing shorted electricity pylons, the main economic cost of the brown tree snake falls on Guam's two main sources of income: tourism and the military. Few tourists will have a close encounter with a snake, but the possibility must deter many potential visitors with even a slight fear of snakes, and clearly birdwatchers needn't bother visiting Guam at all. The cost to Guam of tourists diverted to alternative destinations such as Hawaii must be considerable.

In addition, Guam is the site of a major US military base, and the US Department of Defense is the main source of cargo leaving Guam. A major preoccupation of the military is therefore preventing the brown tree snake from repeating its original stowaway trick and hitching a ride to other, snake-free, islands. The nightmare scenario is the snake reaching Hawaii, where it is estimated it would cost $1.7 billion a year in power cuts alone, plus all the other environmental damage. So far preventing the snake leaving has been completely successful, but searching cargo costs up to $2 million per year.

ZEBRA MUSSEL

Zebra mussels, *Dreissena polymorpha*, are one of evolution's great success stories. So called on account of the stripy pattern on the shell, they are filter-feeding freshwater bivalves that grow to a maximum size of around 5 cm. They are originally native to Russia but the construction of canals and the expansion of commercial boat traffic allowed the species to spread west into most of Europe during the nineteenth century, arriving in both Britain and the Netherlands during the 1820s. They arrived in North America some time in the 1980s, almost certainly as larvae in the ballast water of a ship from Europe, and the first record there is from 1988 in Lake St Clair – the waterbody that connects Lake Huron to Lake Erie. Just one year later they were also in Lake Erie and Lake Ontario, and by 1991 they had colonised all five Great Lakes and the Hudson, Illinois, Mississippi, Tennessee, St Lawrence and Ohio rivers.

Zebra mussels will settle on any suitable (preferably hard) substrate, and that's when the trouble starts. The mussels block pipes that deliver water to cities, factories and power stations, and attach in huge numbers to ship and boat hulls, marine structures and navigational buoys. In the United States they have so far cost the power industry alone billions of dollars, and a recent estimate puts the annual cost to all industries at $2 billion dollars. North Americans often assume that this is a uniquely American problem, but it isn't. Although zebra mussels are a preferred food of many predators, including fish and crustaceans, they grow quickly and produce huge numbers of effectively dispersed larvae, and none of these predators seem capable – anywhere – of having much impact on their numbers. Thus zebra mussels also occur at high densities in Europe and Asia, and cause similar trouble; it's just that on the eastern side of the Atlantic everyone has had much longer to get used to the problem and find countermeasures.

A split section of pipe clogged by an infestation of zebra mussels.

All this would be bad enough, but economic damage is usually susceptible to some kind of technical fix; if we can't outwit an animal with no brain at all, we really aren't trying. For example, zebra mussels don't like to settle on copper pipes, and steel and concrete can be coated with zinc or slippery silicone-based paints. Fairly simple filters can stop the larvae reaching a problem site, and simply upping the flow rate can be effective; zebra mussels only settle in a flow rate of less than 1.5 m sec^{-1}. There are also numerous other promsing counter-measures on trial.

But on top of the economic costs, the zebra mussel is also implicated in major environmental problems. Top of the list is its effect on native American unionid bivalves, or clams. Unionids occur throughout the world, but they are unusually diverse in

North America, with around 270 species. Clams live partly buried in mud or gravel, but the exposed part of the shell is a very attractive place to settle for zebra mussels. By attaching to their shells, zebra mussels can make it more difficult for clams to burrow and move through sediment, and the extra drag can increase the likelihood of clams being dislodged by water movement. In addition, zebra mussel attachment can simply obstruct the openings of clam shells, reducing the ability to feed and reproduce, or preventing the valves from closing. Zebra mussels may also simply compete directly with clams for food, which they are in a good position to do, since they're sitting on top of them.

No one is quite sure if fouling clams' shells or outcompeting them for food is the major problem, but either way zebra mussels have nearly eradicated native clams from America's infested lakes and rivers. In 1998 one observer noted that 60 per cent of native American unionids were endangered and 12 per cent were already extinct. Unionids were in trouble anyway as a result of various human impacts on their habitats, but some researchers estimate that the zebra mussel accelerates the rate of local extinction tenfold. Together with environmental degradation, zebra mussels threaten over 60 species of endemic unionids (i.e. species that occur nowhere else) in the Mississippi River basin with global extinction.

TAMARISK

It is not only non-native animals that we have to worry about. Tamarisks (*Tamarix* spp.) were cultivated in the Americas by the early nineteenth century – the earliest documented record is from the catalogue of Harvard Botanic Garden in 1818 – and by 1868, the US Department of Agriculture (USDA) arboretum in Washington, DC, was growing six species. Gardeners liked the tamarisk's graceful, wispy habit and its fluffy plumes of tiny pink flowers, and it was (and is) widely grown as an ornamental shrub. Its tolerance of drought and salt also attracted attention, and suggested it might be useful for controlling erosion. Others, concerned about what they saw as the 'reckless' growing of wheat in areas that had formerly supported shortgrass prairie (which would later give rise to the 1930s 'dustbowl'), saw a rosy future for tamarisk as a windbreak. Its rapid growth meant that, starting with small cuttings, you could have a functioning windbreak in three years. Planted widely in various habitats, tamarisk quickly made itself at home, the first naturalised stand being noted near Galveston in Texas in 1877. By 1900 it was growing wild in Texas, Arizona, Utah and California.

Tamarisk *is* drought-tolerant, but it enjoys a drink as much as the rest of us, which spelled trouble. The southwest of the United States is dry, and there is much competition for scarce water. The crunch came when the Phelps Dodge mining company wanted to expand its copper mine in Morenci, Arizona, in the 1930s. Mining is a notoriously water-hungry industry, and there wasn't enough water to go round. Casting around for ways to release more water for human use, attention fell on plants that might be wasting water, and in particular on tamarisk, by now widespread along many rivers and across flood-plains. Early studies quickly established that tamarisk had unusually high rates of evapotranspiration (the combination of water loss from the soil surface by evaporation and from plants by transpiration). These were as high as 3-4 m yr^{-1} – crucially higher than the native cottonwoods (*Populus*) that dominated riparian (i.e. riverside) vegetation until tamarisk came along – and suggested that very large quantities of extra water could be made available by removing plants.

One 1952 study estimated that 15 million acres of shrubs (much of it tamarisk, but not all) were wasting up to 25 million

Tamarisk plants thriving along cracked earth in an area that was until recently underwater in Lake Mead, which supplies water to Las Vegas.

acre-feet of water annually across 17 states. If the acre-foot means nothing to you, you're probably not alone – one acre-foot is 1,233,482 litres, or a cube of water a little over 10 m on a side. In other words, a lot of water. Another study reported that a single large tamarisk could get through 760 litres of water a day, and that tamarisk in the south-west used almost twice as much water as all the major cities of southern California. Another way of saying the same thing was that every year, tamarisk was sucking up three times the water used by all the households in Los Angeles.

Wasting water is bad enough, but it's not tamarisk's only offence. Other studies suggested that another effect of tamarisk, and maybe a more serious one, was salinisation, or increasing the saltiness of flood-plains and rivers. Rivers dominated by tamarisk were saltier than those nearby that were still dominated by native willows and cottonwoods. Nature Conservancy guidance put the matter succinctly in 2011:

> Tamarisk is tolerant of highly saline habitats, and it concentrates salts in its leaves. Over time, as leaf litter accumulates under tamarisk plants, the surface soil can become highly saline, thus impeding future colonization by many native plant species.

A popular name for tamarisk in America is saltcedar, reflecting its ability to tolerate (and create) saline soils. As a result tamarisk disrupts the structure of native plant communities, replacing willows, cottonwoods and mesquite by a low-diversity tamarisk monoculture. Dense stands of tamarisk have been condemned as 'biological deserts'.

Finally, there's the knock-on effect on other parts of the ecosystem. In the arid southwest of the USA, riparian zones are crucial habitats for both migratory and resident birds. Riparian woodlands of willows and cottonwoods provide particularly rich habitats for birds, while tamarisk stands are much less useful. In particular, tamarisk is one of the factors thought to be responsible for the decline of the endangered willow flycatcher, and it is recommended that if restoration of bird habitat is the objective, complete removal of tamarisk is the only course.

PURPLE LOOSESTRIFE

While tamarisk was rampaging across the USA's arid south, another very different alien was spreading across the damp north. Purple loosestrife (*Lythrum salicaria*) is a European wetland plant introduced to the USA in the early nineteenth century. It spread steadily, and at first botanists merely noted its presence, but by the 1930s its monopolistic tendencies began to become apparent. In 1940 one observer reported that 'the formerly unique and endemic flora of the [St Lawrence] estuary is being rapidly obliterated by ... purple loosestrife'.

In 1965, managers at the Montezuma National Wildlife Refuge in New York state were trying to create waterfowl habitat by seasonal flooding of a forest floor, but the main beneficiary seems to have been loosestrife, which by 1978 was estimated to make up more than 95 per cent of the plant biomass. In fact, anywhere that periodic reductions of water level were employed to manage vegetation for waterfowl, the main colonist of exposed mudflats was purple loosestrife. Soon loosestrife was estimated to be spreading at 115,000 hectares per year.

Researchers at the Montezuma Refuge set out to quantify the effects of loosestrife, and in particular its effect on native *Typha* species (cattails in America, reedmace or bulrush in Britain). Loosestrife was found to displace cattails, the loss of open water following loosestrife invasion was detrimental to wildlife, and the conclusion was that loosestrife colonisation led to 'serious ecological consequences'. A 1987 review of the spread and impact of loosestrife concluded that 'the replacement of a native wetland community by a monospecific stand of an exotic weed' was 'a local ecological disaster'. In April 2005 it was the USDA Forest Service's 'weed of the week':

It spreads through the vast number of seeds dispersed by wind and water, and vegetatively through underground

*stems at a rate of about one foot per year. Seed banks can
remain viable for twenty years. Purple loosestrife adapts
to natural and disturbed wetlands. As it establishes and
expands, it can outcompete and replace native grasses,
sedges, and other flowering plants that provide a higher
quality source of nutrition for wildlife. The highly invasive
nature of purple loosestrife allows it to form dense,
homogeneous stands that restrict native wetland plant
species, including some federally endangered orchids, and
reduce habitat for waterfowl.*

Various methods of control have been attempted, including
herbicides, flooding, cutting and burning, but none is particularly effective, except where it occurs in small isolated stands,
and in any case the huge seed production and long-lived seed
bank mean it's always likely to return. Biological control has
been attempted by importing four European beetles that eat
the roots, leaves and flowers of loosestrife, but so far they don't
seem to have done much good, either. By 2011 loosestrife
occurred in 48 US states and was costing $45 million per year
in control costs and habitat restoration.

CHAPTER FOUR

GUILTY AS CHARGED?

T he plants and animals described in the last chapter are among the best documented 'invasives' that are seen as threatening local, or indeed national, ecosystems and that have prompted major investment in their control or eradication. But are they guilty as charged, or might they be convenient scapegoats in a more complex picture? Here is some further evidence ...

PURPLE LOOSESTRIFE

Purple loosestrife began to earn a bad name for itself in the USA during the 1930s, although at that stage no one had conducted any research on its actual effects. The first such research was conducted at the Montezuma National Wildlife Refuge in New York State between 1978 and 1980. This research attempted to do several things, including establishing whether loosestrife was excluding *Typha* species (cattails). The

Purple loosestrife – too conspicuous for its own good, especially when it is in bloom, as here in Cooper Marsh Conservation Area, Ontario.

researchers concluded that loosestrife *did* have a negative effect on cattails, but provided no statistics, perhaps wisely since their data show no obvious relationship between the two plants. Two further sets of experiments looked at the effects of muskrats and deer on loosestrife and cattails, but again the results were far from clear. Indeed the muskrat study simply didn't work, since two of three exclosures intended to exclude muskrats failed to do so. Finally they looked at the use of loosestrife and cattails by native birds, but again the work was poorly designed and no statistical analysis was conducted. Nevertheless they felt justified in concluding that loosestrife was a menace.

Very tall people with red hair, big tattoos and conspicuous facial scars rarely have successful careers as bank robbers, and loosestrife has a similar problem: it's just too conspicuous for its own good. Recognition of loosestrife as a problem was based largely on anecdotal observations, which are likely to be particularly unreliable in the case of a tall species with such bright, obvious flowers. This is a well-known problem that standard textbooks warn against: it's easy to conclude that an otherwise rather dull wetland has been completely taken over if you look at it when loosestrife is in flower (especially if this is what you *expect* to find), even if a more careful examination would reveal no such thing.

Frustrated by this unsatisfactory state of affairs, two Canadian researchers set out to conduct a much more thorough study on the Bar River in Ontario. They carefully compared plots with and without various levels of invasion by loose-strife, making sure to include some plots with a dense cover of loosestrife and to match invaded and uninvaded plots as far as possible. Their results allowed them to answer three questions: are plots with loosestrife less diverse than those without, is there any correlation between loosestrife cover and richness of other species, and are there any native species that are less likely to occur in invaded plots? The answer to all three questions was unequivocal: no. In 1999 they published their conclusions:

These results provide no support for the hypothesis that the number of species in wetlands is decreasing in association with the invasion of Lythrum salicaria in Ontario.

Not much has changed since 1999, and a 2010 review by Canadian ecologist Claude Lavoie concluded:

In summary, the image of purple loosestrife depicted by scientific studies (lacking definition) is far removed from that portrayed by newspapers (alarming). Purple loosestrife is certainly an invader, and some native species likely suffer from an invasion, but stating that this plant has 'large negative impacts' on wetlands is probably exaggerated. The most commonly mentioned impact (purple loosestrife crowds out native plants and forms a monoculture) is controversial and has not been observed in nature (with maybe one exception). There is certainly no evidence that purple loosestrife 'kills wetlands' or 'creates biological deserts', as it is repeatedly reported. For instance, 63 insect genera, representing 38 families and seven orders, have been collected from purple loosestrife invaded sites in Manitoba. There are no published studies (at least in peer-reviewed journals) demonstrating that purple loosestrife has an impact on waterfowl or fishes. All the other negative impacts associated with purple loosestrife in the press have not been the object of a study (and many have never even been considered by scientists), except for a possible impact on amphibians that has been tested (to date) only on two species, one reacting negatively.

Which is as close as dry scientific language gets to saying that as far as the evidence goes, persecuting loosestrife is, and always has been, a waste of time.

Even the 'maybe one' example of its ability to crowd out natives and form a monoculture, and the one negative reaction of an amphibian, is doubtful. As far as effects on amphibians are concerned, American toad tadpoles are undoubtedly affected by phenolic compounds leached from purple loosestrife leaves (although other native amphibians are not), but phenol-rich foliage is not the sole prerogative of alien plants; the native swamp loosestrife has exactly the same effect. Invasion of purple loosestrife in North America has tended to consist of a first wave, in which the plant forms very dense stands, but later these stands usually decline, allowing reestablishment of a more diverse community. The only evidence of a strong negative impact of purple loosestrife on plant diversity in the field comes from western North America, where purple loosestrife is a relatively recent colonist. Where loosestrife has been present since the 1930s, for example along the St Lawrence River, dense stands are now rare.

No one has any idea why the ability to form dense stands seems to decline as the invasion develops, but it's interesting to note that purple loosestrife is not an isolated example of this phenomenon. The textbook example is Canadian pondweed, an American native that expanded to plague proportions in the UK and then just as rapidly declined to become relatively uncommon. We don't know what caused the decline in the case of Canadian pondweed, either. But occasionally we do, as in the case of garlic mustard (*Alliaria petiolata*), an aggressive European invader of North American forest understories, much of whose success seems to be due to production of toxic chemicals that negatively affect seed germination and mycorrhizal fungi of native trees. But over time, genetic changes in the plant have led to a marked decline in its toxin production, leading to a major recovery in the performance of native tree seedlings in eastern North America, where it has been present for the longest time.

An irony is that, before loosestrife began to be seen as a problem, American wildlife managers had focused their attention on cattails, which were accused of all the crimes (crowding out other plants, interfering with fishing and boating, and generally being 'aggressive') now laid at the door of loosestrife. Nothing much has changed, except that loosestrife presents a more obvious target.

67

European readers probably deserve an apology at this point. You know purple loosestrife, if you know it at all, as a relatively uncommon and rather attractive marsh plant that is also sometimes grown in gardens, so you're probably wondering what all the fuss is about. To get a feel for the kind of hysteria generated by loosestrife in North America, a good European analogy is probably Himalayan balsam (*Impatiens glandulifera*). Himalayan balsam is seen as a problem throughout northern Europe, is routinely listed as one of the UK's very few really troublesome alien plants, and is on Schedule 9 of the Wildlife and Countryside Act, which forbids its release into the wild. Its misdemeanours are succinctly described by the Rivers and Fisheries Trusts of Scotland as:

Thick monospecific stands shade out low level native plants, reducing diversity and denuding riverbanks of understorey vegetation ... Greater nectar production makes flowers more attractive to bumblebees resulting in less pollination of native species.

Let's take a look at those claims. A big Czech study looked at six different sites, both by comparing invaded and uninvaded vegetation, and by removing balsam and seeing what happened. The study revealed no effects of balsam on the diversity or composition of riverbank vegetation, and the authors concluded that:

I. glandulifera exerts negligible effect on the characteristics of invaded riparian communities, hence it does not represent threat to the plant diversity of invaded areas.

Another study in north-east England found that:

Although Impatiens reduces native species diversity in open and frequently disturbed riparian vegetation, many of the

*species negatively influenced by Impatiens are widespread
ruderal species. Furthermore, management may lead to a
compensatory increase in the abundance of other non-native
species and thus fail to achieve desired conservation goals.*

Which, roughly translated, means that although balsam is a bit
of a thug, the plants that suffer most from its presence tend to
be thugs too, some of them native (e.g. goosegrass, bindweed
and couch grass), but many of them alien (ground elder, sweet
cicely, feverfew and winter heliotrope), and some of these are
just as good at reducing diversity as balsam (or even better).

Himalayan balsam – whose release is forbidden across the UK. Its chief
offence, like loosestrife, may be that, growing to 2 m in height and with
bright purple-pink flowers, it is simply too conspicuous.

A big study of pollinators looked at 14 sites in central Germany and found that balsam is certainly very attractive to bees, but found no effects at all on pollination of native plants, other than a small tendency for native plants growing with balsam to receive *more* visits by honeybees than those in sites without balsam. In fact, if one were to look at Himalayan balsam impartially, it might even be seen as beneficial for native pollinators, since

> the number of flowering plant species during the I. glandulifera flowering peak was lower than before the flowering peak, indicating that I. glandulifera fills a late-seasonal gap in flowering phenology that native pollinators can explore.

In other words, the mass flowering of balsam may give bees a late-season boost, rather in the way that oilseed rape has been shown to do earlier in the season.

Finally, foresters worry about the possible effect of balsam on young trees, but a study revealed that it had no effect on the growth or survival of seedlings of silver birch, Norway spruce or silver fir. The same study did reveal big negative effects on young trees from bramble, but that's native so that's OK.

TAMARISK

Purple loosestrife – a plant that never did anyone any harm and doesn't look like doing so in the future – is easy to defend. Tamarisk is a more complicated customer because, as many people have noticed, bad things do seem to happen around it. But the story is more complicated, for it turns out a lot of the evidence for this is based on old, unreliable data.

There are plenty of things people don't like about tamarisk, but near the top of the list is 'wasting' water. Early studies showed that tamarisk has high rates of water use, and these figures are still quoted today. But modern improved data methods show that tamarisk's water use is actually much lower, and not significantly different from that of native trees like willow and cottonwood. At the larger scale, it's also been shown that evapotranspiration from different rivers is fairly uniform, whatever the amounts of tamarisk.

It may seem 'obvious' that removal of tamarisk (or any other plant that uses groundwater) would increase river flows, but this is surprisingly hard to demonstrate in practice. Even if vegetation removal raises the water table, it's very hard to prevent rapid recolonisation by new plants. One study showed there was a brief saving of water when tamarisk was eliminated, but forecast this would be very short-lived as plants regrew. In another study on the Pecos River in Texas, a large-scale chemical eradication programme killed a lot of tamarisk but had no effect on river flows.

Assuming that, because A follows B, A was caused by B, is a basic error in science. It may simply be coincidence, or – as is likely in the case of tamarisk – both A and B were caused by something else. Tamarisk is accused of creating salty soils and of outcompeting and replacing native plants. But before jumping to either conclusion, one might consider a simpler alternative explanation: tamarisk is better suited to the new environment created by the building of dams, increased water abstraction for human use, altered fire regimes and increased cattle grazing. Tamarisk is certainly salt-tolerant, so when river regulation or irrigation practices raise salinity, it's not really surprising that it moves in. Or that overgrazing leads to the replacement of the more palatable willows and cottonwood by tamarisk. Many studies show that in those few remaining places where rivers flow freely and the water table has not been artificially lowered, the presence of tamarisk does not lead to salinisation, nor does it replace willows and cottonwood.

71

In short the continuing erection of dams, accompanied by the diversion, pumping and redistribution of water, have amounted to a massive habitat creation scheme for tamarisk, and it hardly seems fair to blame the plant for seizing the opportunity.

Tamarisk illustrates another important principle: once an alien invader gets a bad name, it becomes easy to blame it for any perceived environmental problem in the vicinity. The original riparian woodlands were the home of the endangered south-western willow flycatcher, and tamarisk was widely assumed to be one of the reasons for its decline, but it now turns out that in some areas many of the flycatchers nest quite happily in tamarisk, and that fledgling success there is indistinguishable from that in native trees. Indeed, concerned over loss of habitat for the flycatcher, the US Fish and Wildlife Service refused permission for the release of tamarisk biocontrol insects in parts of the bird's range.

Even opinions of exactly the same facts can change when a species falls from grace. As the capacity of Lake McMillan, created by damming the Pecos River in New Mexico, was reduced by sedimentation, tamarisk was at first praised for colonising the inlet delta and preventing sediment from entering the lake. By contrast, tamarisk is now seen as part of the problem for all the usual reasons (see above) and is the target of the optimistically titled Pecos River Basin Water Salvage Project.

Note that, while none of this proves tamarisk is innocent, none of it proves it's guilty either; 'case not proven' is about the most you could say. In the face of such uncertainty, even some of tamarisk's enemies occasionally admit that they're not quite sure why they spend so much time and money worrying about it. Bugwood Wiki, for example, provides in its Invasipedia, authoritative accounts of invasive species from the US Nature

A south-western willow flycatcher, unwitting protector of tamarisk.

Conservancy's Global Invasive Species Team. The site ticks off the usual catalogue of tamarisk's misdeeds:

1) *crowds out native stands of riparian and wetland vegetation*

2) *increases the salinity of surface soil rendering the soil inhospitable to native plant species*

3) *provides generally lower wildlife habitat value than native vegetation*

4) *dries up springs, wetlands, riparian areas and small streams by lowering surface water tables*

5) *widens floodplains by clogging stream channels*

6) *increases sediment deposition due to the abundance of tamarisk stems in dense stands*

7) uses more water than comparable native plant communities

before apologetically concluding: 'However, data to support these claims by various authors do not always exist.'

Tamarisk's real offence – or mistake – is to be found loitering too often at the scene of the crime, the crime in this case being the extravagant use of water by the farmers and citizens of the south-western United States. The various owners of the water in the Colorado River already have legal rights to more water than actually flows in the river, and the gap between that flow and rising demand can only increase. Is it any wonder, trying to square the competing and ulti-mately irreconcilable demands of three million acres of highly productive irrigated farmland, and the needs of the citizens of Los Angeles, Las Vegas, Tucson and Phoenix, that the natural reaction is to look around for someone to blame rather than admitting that, as usual, it's all our fault?

ZEBRA MUSSEL

The zebra mussel is, as we have seen, widely credited with causing severe environmental and economic damage, and, specifically, it is reckoned to be the main threat to the survival of North American freshwater unionid bivalves. Unionids are certainly in big trouble, and several species are already extinct, but their problems began long before the introduction of zebra mussels to America in the 1980s. A familiar cocktail of habitat destruction and degradation (arising from building dams, water abstraction, erosion and eutrophication), pollution, overharvesting and loss of the host fish that parasitic unionid larvae need to complete their life cycles are all implicated. Indeed there's every reason to believe that the decline of unionids would continue even if the zebra mussel had never arrived. As is often the case, blaming an alien for filling the gap left by declining natives looks like shooting the messenger, reflecting the usual unwillingness to recognise that the real problem is us humans, and our unwillingness to take the hard decisions needed to improve things.

The zebra mussel illustrates nicely another odd feature of our attitude to alien species: an unwillingness, and indeed often a blank refusal, to accept that alien species might have any redeeming features whatsoever. As far as aliens are concerned, the balance sheet has only a debit column. The go-to place for everything you ever wanted to know about zebra mussels is the US Army Corps of Engineers Zebra Mussel Research Program, which you can read all about on the website of the Aquatic Nuisance Species Information System. These guys are out to destroy zebra mussels, so anything good they have to say about them can probably be trusted.

So what do they say? Here are a few extracts (lengthy lists of references omitted for clarity):

[Zebra mussels] can have large impacts on benthic [bottom-dwelling] communities, such as increases in benthic plant and algal abundance, increases in the density of benthic invertebrates ... As a result of their filter feeding, Dreissena populations shift suspended matter from the water column to the benthos. As a result, zebra mussels have been associated with increases in water transparency; decreases in turbidity ... After the invasion of Dreissena in Lukomskoe Lake (Belarus), water transparency increased from 1.8 to 4 m, and seston [suspended particulate matter] concentrations decreased threefold.

The large increase in macrophyte [rooted aquatic plants] coverage and biomass in waters infested with zebra mussels is the result of increased water clarity. Improved water transparency allows sunlight to penetrate to deeper levels where macrophytes can now become established. The euphotic [well-lit] depth in Lake St Clair, for example, has now expanded to include most of the benthic surface, and it has now become the host of large areas of macrophyte growth. Increased water clarity in Lukomskoe Lake (Belarus) from 1.8 to 4 m resulted in an expansion of macrophyte cover (from 6 to 30 per cent of total lake area) due to an increase in the depth at which macrophytes can grow (from 2.5 to 5 m).

When zebra mussels enter new habitats, they can quickly become a major component of the diet of molluscivorous fish. This has been documented for freshwater drum in Lake Erie. Utilization of this new food source may lead to increased growth rates and productivity, and this has been documented in studies in the former Soviet Union. In general, there is an enhancement of all benthic feeding fishes, even those that do not feed on Dreissena, because zebra mussels increase the biomass of other benthic invertebrates.

Zebra mussels can be a valuable food source for waterfowl species. They are now the dominant food consumed by lesser scaup and Goldeneye in western Lake Erie and northern Lake St Clair. Their consumption by migrating or overwintering birds is well-documented. Zebra mussels are the main food item for 310,000 overwintering diving ducks in Lake IJsselmeer in the Netherlands and a frequent prey item for diving ducks in Lake Erie during their fall migration. Because of their importance as a prey item, dramatic increases in flock sizes can occur following zebra mussel colonization of a water body, as observed in Lake Erie, Germany, and Switzerland.

The location of zebra mussel populations in Europe cannot only affect waterfowl distribution, but also the timing and routes of their migration. The geographical range of tufted duck in England has expanded due in part to the spread of zebra mussels. Food abundance and availability, particularly of Dreissena, were viewed to be the main factors governing lake choice by overwintering diving ducks in Switzerland.

Soon after the arrival of zebra mussels in western Lake Constance in the late 1960s, 45,000 tufted duck, pochard, and coot were observed overwintering; this represented a ten-to-fiftyfold increase over previous levels.

Let's not get carried away here; I'm not suggesting that the world is in every respect a better place for having zebra mussels in it. But clearly there are significant items on the credit side of the zebra mussel balance sheet, which are routinely ignored by its detractors. It's hard to spin increased water clarity and more macrophytes, invertebrates, fish and wildfowl (and, although not mentioned in the above quotes, more crayfish too) as undesirable, but that doesn't stop people trying. For example, increased macrophyte growth very occasionally leads to more vegetation washed up on beaches, reducing their recreational use. But in reality, eutrophication of freshwaters by erosion and fertiliser runoff, leading to algal blooms and increased turbidity,

Ducks in Lake Constance – beneficiaries of rich stocks of zebra mussels.

is a global problem, with reduced macrophyte growth one of its most obvious symptoms. Here's Wikipedia on the subject of macrophytes:

> *In lakes macrophytes provide cover for fish and substrate for aquatic invertebrates, produce oxygen, and act as food for some fish and wildlife. A decline in a macrophyte population may indicate water quality problems.*

In truth the ability of zebra mussels to filter particulate matter from water, and thus return the murky soup produced by human activity to something like its natural state, is remarkable. If zebra mussels were native, there's every reason to expect they would be hailed as environmental heroes, rather than vilified as public enemy number one. Of course, like all filter feeders, they sometimes accumulate toxic amounts of pesticides or heavy metals, which can then be passed on to the organisms that feed on them. But blaming the poor old zebra mussel for this (and trust me, we do) is bizarre – *we* dumped the toxins in the water. The zebra mussel's extraordinary abilities can even be put to use in industrial filtration and water treatment, so even its undoubted economic effects are not all one-way.

A postscript to the zebra mussel story is the round goby, a small fish that arrived from the Black and Caspian seas in 1990 and has spread rapidly throughout the Great Lakes. The round goby has given rise to the usual panic, especially on account of its voracious appetite, but this appetite is mainly for zebra mussels, which make up most of its diet. In fact, the round goby is an excellent converter of mussels into food for larger predatory fish like walleye, and also now makes up more than 90 per cent of the diet of the threatened Lake Erie water snake. Of course, it gets the blame for passing on the toxins and heavy metals accumulated by zebra mussels to larger fish, but it's no more to blame for this than the mussels are.

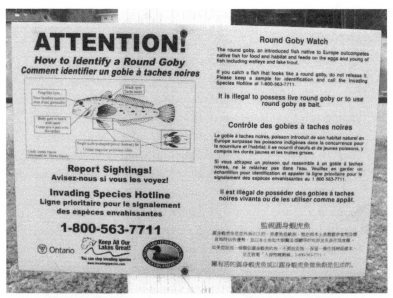

Round goby alert in Ontario. Watch out for those 'frog-like eyes'.

Like all invasions, the zebra mussel story is a dynamic one, with native predators adapting to the abundant new food source. Zebra mussels have become an important food for North American predators from turtles to crayfish, and now constitute a large proportion of the diet of whitefish, the most important commercial fish in the Great Lakes. Unsurprisingly, experiments show that mussels suffered more predation when exposed to fish predation than in cages protected from fish, but exposed mussels disappeared faster in 1998 (seven years after the invasion) than in 1994 (three years after the invasion); the fish are getting *better* at exploiting zebra mussels. Native blue crabs also consume large numbers of zebra mussels, and they too seem to be getting better at exploiting this new food source. Because crabs prefer large mussels, there's also been a recovery in large zooplankton such as rotifers, which are too large to be filtered by small mussels.

OK, BUT WHAT ABOUT THE BROWN TREE SNAKE?

Everyone agrees that islands are uniquely susceptible to invasion by alien species (although hard evidence is surprisingly hard to come by). But what is not in doubt is that islands are usually short of predators, and that remote islands generally lack mammalian predators completely. Unfamiliarity of island animals with predators, along with the tendency of island birds to evolve flightlessness, means that introduced predators, especially rats and cats, have caused mayhem on islands throughout the world. Guam is a classic example, the only unusual thing being that the predator in this case is a snake. But there is no defence here: Guam's brown tree snake is a Bad Thing.

However, we are in danger of using the brown tree snake and its ilk to tar all aliens with the same brush, when the overwhelming majority do no harm at all. In fact, my detailed examination of four well-documented villains, leading to the conclusion that only one of them is proven to be as bad as most people believe, almost certainly exaggerates the problem; it's highly unlikely that one in four introduced species causes serious harm. And if you harbour the suspicion that I have deliberately picked on four inoffensive straw men to demolish, nothing could be further from the truth: all four of the species considered here appear regularly on lists of the world's worst invasive species. Tamarisk, for example, has been described as the 'second-worst invasive plant species in the United States'.

The only unusual thing my four chosen species have in common is that someone has taken the trouble to subject their alleged crimes to detailed scrutiny.

CHAPTER FIVE

IF IT'S NICE, IT MUST BE NATIVE

You will have gathered by now that some alien plants and animals are deemed to be very bad news indeed, even if the evidence for such bad behaviour isn't always all it seems. Thus, given the amount of time and vast sums expended on containment and attempted eradication, it would be reassuring to think that 'native' and 'alien' were nice, unambiguous categories, and that we were always sure which species belonged in which category. Sadly, as I have already pointed out, we aren't very good at this at all.

THE NATIVE BRITISH FLORA

As we saw in Chapter 1, many apparently native British plants are in fact archaeophytes, i.e. ancient introductions. But that turns out to be only half the story. If we exclude the numerous 'microspecies' of brambles, hawkweeds and such, which are recognisable by, and of interest to, only a tiny minority of dedicated specialists, the consensus is that Britain has around 1,500 native species of higher plants. By the standards of global, and especially tropical, biodiversity this is a tiny number – Britain is very far from a biodiversity hotspot. At the same time, Britain has large numbers of naturalists, both amateur and professional, who have been busy mapping, counting and cataloguing the flora for centuries. Accordingly it's fair to say that the British flora is the most thoroughly studied on the planet and that, for example, no native species remains to be discovered – something that cannot be said for the rest of the globe. (For the collector of obscure facts, the last unambiguously native plant species to be found in the UK was Iceland purslane, *Koenigia islandica*, first collected in 1934, but not correctly identified until 1950. *Koenigia* is an Alpine and both very rare and very small, in fact more or less invisible unless you're on your hands and knees, which accounts for its very late discovery.)

I mention this to emphasise that if British botanists don't know what's native and what's alien, there's little hope for the rest of the world, which generally has many more plants and far fewer botanists to count them. Even in Britain, we can still talk only of 'around' 1,500 native plants, because included in that list are a surprisingly large number of doubtful cases. In fact around 20 per cent of Britain's arguably 'native' flora (about 300 species) are more accurately described as 'doubtfully native', meaning no one is quite sure. If you ask a botanist to swear on his copy of *Stace* (the standard British Flora) that one of these plants really is native, only a small minority of incurable optimists will cheerfully do so.

Roughly half of these doubtful cases are archaeophytes, which you might think is fair enough; these are species

A suspect British native – the attractive and easily recognised snake's head fritillary, *Fritillaria meleagris*.

introduced before 1500, which means we can't expect to find any direct documentary evidence of their introduction. But after 1500, surely we are on firmer ground? Not necessarily. David Pearman, ex-president of the Botanical Society of the British Isles (BSBI) and all-round iconoclast, has now assembled all the available evidence on these 150 suspect plants and come to some surprising conclusions. As an illustration, let's look at the snake's head fritillary, *Fritillaria meleagris*, a plant that generations of gardeners and botanists (including me) have been brought up to believe is indisputably a British native, because that's what all the books say.

The least we would expect of a genuine native is that it is unquestionably native just across the English Channel, in the territory it must have crossed on its way to Britain. That's because to be officially native, a species must not only have reached the UK on its own, but it must have done so *starting from* somewhere it is also native. Thus if an Asian or North American species is introduced to mainland Europe, and then makes its own way to the UK, it's still an alien. *Fritillaria* fails this test spectacularly; it's not found at all in north-east France and is regarded as an introduction in Germany. In Denmark it's also regarded as an alien, and they even have a date (1647) for its introduction. In Sweden they have not only a date of introduction (1658, and possibly earlier) but also a place, Uppsala Botanic Garden. It didn't escape from there into the wild until 1742. The nearest place where *Fritillaria* seems to be genuinely native, or at least regarded as such with any conviction, is Poland.

The history of *Fritillaria* in Britain is equally suspicious. It was certainly in cultivation in Britain by 1597, and possibly by 1578 (plant names hadn't really settled down that long ago, and it's sometimes not clear exactly which plant is being talked about). On the other hand, the first record in the wild is 1736, and even that is an outlier; no one admits to seeing it again until 1776. That's a *very* late date for a real British native, especially such a colourful, unmistakable and attractive one. In other words, if you believe *Fritillaria* is native, you have to assume that generations of seventeenth- and eighteenth-century botanists conspired not to mention it, a plot rivalled only by NASA faking the American moon landings, presumably on their day off from guarding the alien artefacts at Roswell.

By applying similar reasoning, Pearman has concluded that at least three-quarters of the doubtful cases are almost certainly or very probably introductions. Not only that, but quite *recent* introductions – remember, these are the species left over after we take out the known (or suspected) archaeophytes.

Of course, you can never be quite certain about any of these plants, which is why they are in the doubtful camp in the first place, but we can say for sure that *none* is a guaranteed native. Some of the early introductions were accidental, but most of the later ones are escapees from gardens.

Which leaves us with the question: why are so many people willing to believe that so many plants are native, when the evidence, viewed objectively, is so strongly against such a conclusion? One clue is the sort of plants we're talking about; not just fritillaries, but cypress spurge (*Euphorbia cyparissias*), gooseberry (*Ribes uva-crispa*), monkshood (*Aconitum napellus*), fly honeysuckle (*Lonicera xylosteum*), snowdrop (*Galanthus nivalis*), summer snowflake (*Leucojum aestivum*), grape hyacinth (*Muscari neglectum*) and mezereon (*Daphne mezereum*), among others.

These are harmless, attractive, often rare plants that you *want* to be native, and therefore if you find them growing apparently wild, you're likely to assume they are. When British wildlife charity Plantlife asked its members to vote for 'county flowers', *Fritillaria meleagris* came out top in Oxfordshire. You wouldn't want to be told that the plant thus honoured was really an illegal immigrant, would you? Not that the citizens of Oxfordshire were alone in being seduced by the charms of a foreign flower; Tyne and Wear chose monkey flower (*Mimulus guttatus*) from America, Lancashire chose *Rosa gallica* (the 'Red Rose of Lancaster') from central Europe, and Nottinghamshire settled on autumn crocus (*Crocus nudiflorus*) from the Pyrenees. Nor are the British unique in this respect; many US states have non-native state flowers, and nearly a third have the (non-native) honeybee as their state insect.

So lots of plants that most people probably think are British natives (if they ever think about it at all) are not. Do we care? Not at all, apparently – like the people of Oxford, if we like a plant we don't even stop to think whether it's native or alien. For example, *Plantlife* has this to say about British cornfield weeds:

From the iconic cornflower, golden corn marigold and brilliant red poppies, the wild flowers of this intensive landscape do their best to provide diversity, colour and beauty.

Most British cornfield weeds, including those three mentioned, are archaeophytes (ancient human introductions), but these are attractive and now declining plants and *diversity, colour and beauty* easily trump their non-native status. The county plants voters think so, too – both Essex and Norfolk elected the corn poppy (*Papaver rhoeas*) as their county flower. It's only when we've already decided for other reasons that we don't like a plant that we add being foreign to its list of misdemeanours. You can bet that if any of these formerly feared weeds was still capable of seriously annoying us, we would be less keen on overlooking its history. For example, Britain's worst alien plant, if we measure that purely by economic impact, is the wild oat *Avena fatua*, which has been present since the Bronze Age (3,000 years ago), but is still regarded as an undesirable alien. But then we like poppies, and we don't like wild oats.

HARES, RABBIT AND CRAYFISH

Are the British like this only about plants? No, the same lopsided criteria are applied to animals. For example, nobody much likes wild rabbits (an introduction, but one that *was* present in Britain before the most recent glaciation), but we appear to have a soft spot for the brown hare (also native in a previous interglacial, but introduced in this one). In fact hares now have their very own Biodiversity Action Plan (BAP), a government instrument designed to deliver our commitments to the UN Convention on Biological Diversity. Hares have declined in Britain in the last 50 years, owing to the same catalogue of land use changes (essentially an increase in the intensity of farming) that have severely affected farmland birds, butterflies and many other species. The BAP advises various actions to halt the decline in the population of hares. Does it matter that hares aren't native? Apparently not.

Generally speaking, we like animals and plants that are attractive and don't cause any trouble; indeed the species we really like are those that are in decline, and we often wade in on their behalf. Hence no one likes (successful) rabbits, but everyone likes (threatened) hares. We don't seem to mind that hares are introduced, though we're much happier trying to help native species, and indeed as I noted in Chapter 2 we seem to assume that native species somehow *ought* to be here and therefore *deserve* our help in a way that aliens do not. So powerful is this need to believe that the objects of our affection are native that it can even affect our judgement of whether a species *is* native.

The white-clawed crayfish (*Austropotamobius pallipes*) provides a perfect example. This freshwater crustacean has declined dramatically in Britain (and in some other European countries) in recent years, a decline largely attributed to competition from the American signal crayfish, and to crayfish plague which is carried by the American species.

We want to protect 'our' crayfish from the aggressive, American, disease-carrying interloper. But is our crayfish native? Well, there is no written reference to crayfish in England either before or during the Middle Ages, nor have crayfish remains been found in middens from that time. This is surprising, since crayfish were legally fish and thus extremely popular with monks and nuns, who ate them in large numbers at times when their religion forbade meat. The habit soon spread outside

White-clawed crayfish – a German import under threat?

monasteries, and the earliest German cookbooks are full of recipes for crayfish. On the other hand, one of the first written references to crayfish in England is from 1586, and is a record of their introduction by Sir Christopher Metcalfe to the River Ure in Yorkshire from 'the south part of England'. Crayfish were a valuable food, hence their translocation from one end of the country to the other. But why didn't this happen earlier? Together with the absence of any earlier written or physical evidence, the obvious inference is that crayfish were introduced to England from Europe around 1500.

The genetic evidence is consistent with this interpretation: French crayfish are genetically diverse, but British crayfish are not, and are genetically identical to crayfish from northern France. Introduction of crayfish to Scotland was more recent still, only in the last 150 years, so Scottish crayfish 'therefore cannot be considered as indigenous'. Of course, you'll not be surprised to hear that that doesn't stop Scotland imagining that they are; the signal crayfish is blamed in Scotland for 'eradication of indigenous population of white-clawed crayfish *Austropotamobius pallipes* through direct competition and transmission of lethal crayfish plague' by the website invasivespecies-scotland, even though Scotland doesn't have an indigenous crayfish.

You may be feeling a bit confused by this whole business of 'nativeness' by now, and if you're not, you certainly ought to be. Crayfish have been in England for around 600 years, which is apparently long enough to make them native (like hares, crayfish have their own BAP, and are routinely described by all the relevant conservation agencies as native). But crayfish are not native to Scotland, because 150 years isn't long enough. Recall *Fritillaria meleagris*, which we now realise is an alien. *Fritillaria* was in England by 1578, which seems a long time ago (almost in crayfish territory), but that was only in cultivation, so was it *really* here? Did it only start earning its citizenship in 1736, when it was first recorded in the wild? And where does that leave *Rhododendron ponticum*, widely regarded as Britain's worst environmental weed, first introduced to Britain in 1753?

To me, 1753 doesn't seem all that different from 1736, but is *Rhododendron* just too nasty ever to earn enough brownie points to become an honorary native? Two factors complicate that question. If we go back far enough, *Rhododendron ponticum* was formerly much more widely distributed, and was certainly present in the British Isles in the interglacial before the present one. Some would regard that as the ultimate prior claim to native status. In any case, the plant that is currently causing so much trouble in Britain and Ireland is a hybrid of *R. ponticum*, *R. catawbiense* and *R. maximum*, and perhaps *R. macrophyllum*. If we accept the proposed name of *R.* × *superponticum* for this new plant, what is *its* status? It has a good claim to be a British

89

native, since it evolved in Britain and is found nowhere else, even though all its parents are aliens. Or is it that rare thing, a stateless plant?

BEAVERS IN BRITAIN

So elastic is the concept of nativeness that it can even be revoked, if someone thinks the native in question might be an inconvenience. The European beaver, *Castor fiber*, is a British native, but absent since it was hunted to extinction in the seventeenth century. After much debate and legal argument, a trial reintroduction was begun in 2009 at a remote site in Argyll, Scotland. But beavers (of unknown origin) had also been doing quite nicely on the River Tay in central Scotland for the last ten years. Scottish Land and Estates, the body that represents Scottish rural landowners, adopted what it describes a 'robust attitude' to these Tayside beavers; in other words, it wanted them exterminated. It's quite hard to fathom the reasons for this, but it seems to stem partly from a belief that beavers might interfere with migrating salmon, something for which there is not a shred of evidence (indeed, beavers seem to be good for salmon). To shore up their case, Scottish Land and Estates maintain that beavers aren't 'proper' natives, and that (as reported in the *Independent* newspaper) 'the beaver is in no sense a Scottish resident, since the landscape has, it contends, changed out of recognition since the animal died out 400 years ago'. Note what a weasely idea this is, managing to hint that the extinction of British beavers was both natural and inevitable, and that the modern landscape is somehow unsuitable for beavers. In other words, that the thriving River Tay beavers need to be shot for their own good.

The Tay beavers are also apparently objectionable on genetic grounds. European beavers occur as far east as

Tayside beavers – mother and child. Cute, but not true Scots.

Mongolia, but the official position is that Norwegian or at least western beavers are likely to be closest genetically to the extinct UK beaver, so only they should be introduced. But all current western beavers are derived from tiny remnant populations and are genetically very depauperate, raising, as wildlife biologist Duncan J. Halley states, 'the animal welfare implications of reintroducing inbred animals likely to suffer from a heightened level of genetic abnormalities than would a stock of mixed western or mixed western and eastern origin'. Introducing a mixed eastern/western population would provide maximum genetic diversity and 'would have no impact on the wider ecological role of the species, as behaviour appears identical regardless of origin'.

Although the exact origins of the Tay beavers are unknown the likely source is captive animals in the area, which are of mixed western and eastern origins, descended from a robust hybrid used for reintroduction to Bavaria. In other words, if a successful population of adaptable, healthy beavers is the primary aim, there is a powerful argument for letting the Tay beavers get on with it.

THE MISUNDERSTOOD DINGO

My examples so far are all British. So is it just Brits that are uniquely paranoid about origins, as befits a nation still obsessed by subtle distinctions of class? It's a nice idea, but all the evidence suggests that an irrational attitude to aliens is a global problem. Consider the Australian view of the dingo, an animal with a bad name and few friends, despite being probably the best friend Australian conservationists have. To understand why, we have to delve into some history, and a little ecology.

When it comes to mammal extinctions, Australia is a global accident blackspot. Since European settlement, 18 species of mammals have become extinct there – almost half the global total over the same period. There's little doubt that the main cause has been introduced predators, mainly foxes and the feral offspring of domestic cats. For small ground-dwelling marsupials, foxes and cats are the stuff nightmares are made of. Such middle-sized predators have benefited from what ecologists call 'mesopredator release', which is basically 'when the cat's away the mice will play', only moved up a trophic level. Throughout the world, wherever large top predators are eradicated or reduced in numbers, the result is an explosion in the population of smaller 'mesopredators', usually with catastrophic consequences for the prey of these smaller predators. Australia is no different, and there is plenty of evidence that where dingoes are abundant, cats and foxes are rare. Foxes in particular both fear and avoid dingoes. The result is that the decline of small marsupials is closely linked to the density of dingoes; the more dingoes, the better the small native mammals are doing. Where dingoes are rare, cats and foxes are able to cut a swathe through the local dunnarts, bilbys, potoroos and rock wallabies with impunity.

Since the last thylacine died in Hobart Zoo on 7 September 1936, and the Tasmanian devil is now confined to Tasmania,

Australia's useful but unloved predator, the dingo.

the dingo, *Canis lupus dingo*, is the largest surviving predator in Australia. Unfortunately it has always been disliked and persecuted, especially in sheep country, and it's now also threatened by hybridisation with domestic dogs. Its chief mistake, however, is not to be entirely native. Dingoes arrived in Australia about 4,000 years ago, brought by seafarers from south-east Asia. And yet this foreign interloper holds the key to the survival of many of Australia's threatened small marsupials. This has finally been recognised – although maybe too late – by the state of Victoria, which has moved to list the dingo as a threatened species.

The Melbourne *Herald Sun* explained the matter well to its readers in 2008:

> *Dingoes will be listed as a threatened species in Victoria under a state government move to protect the animal from extinction. Breeding with wild dogs and habitat destruction has decimated dingo populations in Victoria, with the dog only surviving in two isolated pockets of the state. Pure bred*

dingoes are found in the mountains in the state's east and in national parks in the Mallee region. The Department of Sustainability and Environment will establish an action plan to protect the dingo, which is likely to include strategies to control wild dog populations which are the main threat to the pure dingo.

DSE biodiversity and ecosystem services executive director Kimberley Dripps said dingoes used to be widespread across Victoria but were now only found in small numbers. 'The dingo has a role in balancing the ecosystem. They do eat kangaroos and wallabies, but they are also effective in reducing the numbers of red foxes and feral cats which are pests,' she said. 'We need the top order predators to balance the ecosystem.'

The dingo's listing as a threatened species comes after Victoria's Scientific Advisory Committee warned the animal was close to extinction. Environment Minister Gavin Jennings said the dingo would now be protected under the Flora and Fauna Guarantee Act 1988. 'We're working to protect the pure Dingo while still supporting appropriate control of wild dogs, hybrids and pure Dingoes which threaten livestock,' he said. 'It is vital that we do what we can to ensure the survival of this threatened species, which is one of Australia's few indigenous, top-order predators. Dingoes have been part of the Australian ecosystem for thousands of years and have an important ecological role.'

Victoria's belated decision represents a rare outbreak of common sense.

CARIBBEAN RACCOONS

The dingo is a nice illustration of an animal that, by the usual criteria (elastic and ill defined as they are), deserves to be recognised as a native, and probably would be if we hadn't decided that we don't like them. The other side of the coin is species that clearly are not native, but where we have decided (again for no very good reason) that we do like them, and therefore we want them to be native. For a fine example, look no further than raccoons.

Raccoons (*Procyon* spp.) are found throughout the Americas, from Canada to Amazonia. Two species are widely distributed, the northern raccoon (*P. lotor*) of northern and central America, and the crab-eating raccoon (*P. cancrivorus*) of central and South America. At various times, raccoons on Caribbean islands have also been recognised as separate species, including those on Barbados, Guadeloupe and New Providence in the Bahamas.

An examination of the evidence, however, demonstrates beyond any doubt that these extra raccoon species are the purest wishful thinking. For a start, the islands concerned have no other non-flying mammals, apart from a single endemic rodent in the Bahamas. This would make the supposedly endemic raccoons almost unique and unusually lucky, not once but on at least three occasions. Nor does the genetic or morphological evidence support their endemic status; indeed, the skulls and skins used to describe the island species all turn out to be immature or otherwise unusual individuals. There are no archaeological or palaeontological records of raccoons on the islands, and the people who were around at the time were sure they were introduced. For example, German naturalist Johann Schöpf had this to say after visiting the Bahamas in 1784:

Of wild quadrupeds, there are but two species, properly only one, indigenous to these islands. The Racoon is found only on Providence Island, of which it is no more originally a native than the rats and mice brought in by ships. From one or more tame pairs of these droll beasts, brought over by the curious from the mainland, and afterwards escaped by chance in

the woods, the race has amazingly increased, to the great
vexation and damage of the inhabitants, who can scarcely
protect their house-fowls from these stealthy thieves.

All three 'endemic' raccoons are in fact recent introductions from North America, and are identical to *P. lotor* in
every respect. The citizens of Guadeloupe were mortified
by this revision, since 'their' raccoon had been adopted as a
flagship species by local conservationists, and even chosen as
the emblematic species of the Parc National de Guadeloupe.
But in the Bahamas, where they still shared Schöpf's original
opinion, they were delighted. As Don Wilson, a taxonomist
at the National Museum of Natural History in Washington,
DC, reports: 'They instantly changed the raccoon's status from
endangered to invasive species and set up a control programme
to eradicate them.' The status of the Barbados raccoon is
academic, since it is recently extinct, but at least we now
know it's not yet another extinct species. But note that on no
island has anything actually changed; raccoons may have been
protected endemics yesterday and undesirable aliens today, but
they're still the same raccoons, as cuddly or as troublesome as
they ever were.

THE TANGLED TALE OF THE POOL FROG

Excellent as dingoes and raccoons are as examples of the ability of the human brain to entertain simultaneously almost any number of mutually incompatible ideas, they are easily outshone by another British example: the tangled tale of the pool frog.

Britain is not well endowed with amphibians, and it's generally assumed there is only one native frog, the common and widely distributed *Rana temporaria*. In mainland Europe there are many other frogs, and one of them, the pool frog (*Rana* – now *Pelophylax* – *lessonae*) has been repeatedly introduced during the last 200 years. Most of these introductions have come to nothing, but a few populations survive. Recently, the possibility has been considered that the pool frog may have been here before these introductions, and is thus an overlooked native. (The introduced British frogs came from France, but there are also very small populations in Sweden and Norway.)

On circumstantial grounds, researchers suspected that one particular population of pool frogs might be the elusive native. This population was very small and very endangered – so endangered, in fact, that the researchers managed to experience the very moment of its extinction, succeeding in capturing the last surviving (male) individual. DNA from this one frog, together with that from four museum specimens from the same area, also allegedly native, were compared with DNA from frogs from mainland Europe, from Scandinavia, and from British frogs known to have been introduced. The (perhaps) native British frogs turned out to be most like those from Scandinavia.

At the same time, examination of tens of thousands of subfossil frog bones revealed two pelvic bones, both around 1,000 years old, that appeared to be from pool frogs, thus predating any known introduction. It's worth noting, however, that the use of the same diagnostic criteria on a collection of *known* pool frog bones from Germany had 'a success rate (after rejecting ambiguous specimens) of 93 per cent', which doesn't fill one with confidence.

Other evidence was also a bit mixed. British ('native') frogs have calls more like those from Scandinavia than those from France, but written evidence of pool frogs in Britain before 1800 is nonexistent; if they were present, no one noticed them. Nevertheless, impressed by the genetic evidence, and the absence of any evidence (and indeed any likelihood) of introduction of frogs to Britain from Scandinavia, the researchers concluded that the pool frog is indeed a British native. British frogs are thus interpreted as part of a northern race of pool frogs that spread north at the end of the last glaciation, some carrying on into Scandinavia, others turning left across the (then existing) land bridge into Britain.

This naturally threw the conservation community into a panic. Suddenly, pool frogs were no longer a minor herpetological curiosity, they were Britain's most endangered amphibian. Indeed, with the death of the lone male, not just endangered, but actually extinct. What to do? The answer was to 'reintroduce' pool frogs from Sweden, together with a programme of tree clearing, installation of artificial water supplies, dredging of existing ponds and creation of new ones, all designed to create the perfect habitat for pool frogs, followed by monitoring to see how they were getting on.

The legal aspect is perhaps even more interesting, since pool frogs went from having no friends and no rights, to having a surfeit of both. The new frogs had, of course, to be legally protected, but this protection does not extend to the existing populations of French frogs. Or, as Natural England put it, the 'unauthorised' frogs from France are 'of no conservation importance'. Not that you or I, or indeed anyone except a tiny handful of experts, could hope to distinguish Swedish pool frogs from French ones. In fact separating pool frogs from the native common frog would tax most of us.

Just to complete the picture, it's worth mentioning that pool frogs themselves are a bit of a moveable feast. The pool frog (*Rana lessonae*) is widespread in mainland Europe, as are its close relatives the marsh frog *R. ridibunda* and the edible frog *R. esculenta*. All three frogs are 'morphologically very variable and are often difficult to distinguish in the field' (I quote here the experts who were so certain about two 1,000-year old frog bones). They also hybridise, and in fact the edible frog is a fertile hybrid of marsh and pool frogs; until the 1970s, pool and edible frogs were generally considered to be subspecies of the same species (*Rana esculenta*). All three frogs have been introduced from Europe on many occasions and populations of all three survive; if the 'native' pool frog is successful, it is likely to meet and hybridise with all of them.

It's worth pausing at this point and taking stock. Leaving aside for the moment that you would need a very good lawyer if you wanted the evidence for native British pool frogs to stand up in court, 'nativeness' is here being used to distinguish frogs that hardly anyone (and certainly not the frogs themselves) could tell apart. As the researchers themselves say, if the pool frog is native, it 'would be Britain's most endangered amphibian, rather than an alien which might be ignored or even extirpated'. The *best* an alien pool frog can hope for is to be ignored, the worst is to be hunted down and eradicated, while the full force of the law will fall on anyone who harms the 'native' frog or damages its habitat. Yet the 'native' tag has changed nothing; no one is suggesting the native frog is less ugly, nor likely to play any different role in the ecosystem. In what way are we, or even the frogs, any better off than we were before?

NATIVENESS UNDER ATTACK

The more you look at it, the harder it becomes to defend nativeness against determined attack. For one thing, as I pointed out in Chapter 2, it's a relatively modern obsession. Another problem is its fragility: once it is tainted by human agency, nativeness, like virginity, is gone forever. And yet, despite nativeness now deciding the absolute superiority of one raccoon (or frog) over another, it's a criterion that is often suspended for no obvious reason. When adding up all the evil done by alien species on the negative side of the balance sheet, there is never the opposite exercise, adding up the crops, livestock and other useful species that have been introduced. Yet usefulness surely trumps non-nativeness.

Where, for example, would the Wild West be without horses, even though horses were introduced to America by the Spanish in historical times? Curiously, horses are now widely regarded as alien in the USA, even though nearly all of horse evolution took place there and they only became extinct as recently as 8,000 years ago. How many people could live in Australia if they were stuck with only the animals and plants that are native there? Not many, is the answer. But if a formerly useful alien, such as the rabbit in Britain, starts to show signs of making itself too much at home, it quickly finds itself once more on the debit side of the balance sheet.

Moreover, it's not the *fact* of human intervention that removes nativeness forever, it's the timing and scale of human intervention. In Britain, no one is quite sure if you have to be around before the Neolithic, or before the Romans, or even the Normans, to be native. If you're inoffensive (and threatened) enough, like the white-clawed crayfish, before 1500 seems to be good enough. In America, the modern interpretation of 'native' seems to mean before 1492, even though pre-Columbian Americans domesticated plants and indulged

100

in long-distance trade. In Australia, beating Joseph Banks to landfall makes you native, despite plenty of evidence that plants and animals were moved around the wider region for millennia before the *Endeavour* pitched up in Botany Bay in 1770.

In other words, pre-technological, pre-European societies, however advanced, and whatever they may have got up to, are often – but not always (as with the dingo) – exempt from the definition of human agency. It's not always easy to tell, of course, since the definition of nativeness is frequently adjusted ad hoc to help to prop up human preferences decided on the basis of other criteria. Just as well, too, since a too strict application of nativeness would require the return of Europe to a pre-Neolithic ecological state.

CHAPTER SIX

A SHORT COURSE IN ECOLOGY

S o where does that leave us? We now know that we (and by 'we' I mean everyone, but especially ecologists like me) often don't know what's alien and what's native, and sometimes we don't even know what we mean by native and alien. And even when we're sure that something is alien, we often aren't sure whether it's a problem or not (or more often we *are* sure, but we don't know *why* we're sure). You could therefore be forgiven for wondering what else we don't understand, and the answer is quite a lot.

SOME NICHE THEORIES

To understand the problem, it's oddly instructive to begin with the basics, which in the UK means the biology syllabus for 16-18 year old children. I can't start with ecology for younger children, because the subject scarcely exists before we get to age sixteen. And even here, in the 77-page syllabus, ecology gets one sparsely populated page, 120 words in total. This stipulates five learning outcomes, of which the first is: define the terms *habitat, niche, population, community* and *ecosystem* and state examples of each. The key word here is *niche*, which is central to our understanding of how *communities* work.

Some definitions (from the 16-18 school syllabus):

NICHE: *the functional role or place of a species of organism within an ecosystem, including interactions with other organisms (such as feeding interactions), habitat, life-cycle and location, adding up to a description of the specific environmental features to which the species is well adapted.*

COMMUNITY: *all of the populations of all of the different species within a specified area at a particular time.*

So even at this basic level, where ecology is covered in only the most abbreviated form, the niche is a key concept. Further up the academic ladder, things are exactly the same. In fact, ever since the term was coined by Joseph Grinnell in 1917, niches have been the principal tool to explain the composition of plant and animal communities, and specifically questions like why there are so many different types of organisms in one habitat, and so few in another, and why different communities (in different habitats) contain such different kinds of species. In other words, all the classic questions about biological diversity.

Here's a widely used undergraduate biology textbook:

If an organism's habitat is its address, the niche is its occupation. Put another way, an organism's niche is its ecological role – how it 'fits into' an ecosystem. The niche of a population of tropical tree lizards, for example, consists

*of, among many other variables, the temperature range it
tolerates, the size of the trees upon which it perches, the time of
day in which it is active, and the size and type of insect it eats.*

Niches are important for biodiversity because any habitat contains only so many; tree lizards may be active at different times, in different parts of the canopy, and specialise on different foods, but there is a limit to all this, and when all the possible variations have been exploited there's no room for any more lizards. Crucially, two lizard species that are too similar cannot coexist, an idea so important in ecology that it has a name: the *competitive exclusion principle*.

A nice analogy is the shops a single high street can support. One baker is fine, but a second may have difficulty establishing. To some extent, bakers (and lizards) can get round this by specialising and thus dividing up the available resource (food for lizards, customers for bakers), but again there is a limit. Ecologists have a name for this, too: the niche of an organism under ideal conditions, in the absence of any external constraints, is its *fundamental niche*. But in reality, an organism usually has to put up with all kinds of competitors, predators and other problems, and the narrower range of conditions it can exploit under those constraints is its *realised niche*. The concept is familiar to gardeners: most plants can be grown in soils and climates that they are unable to exploit in the wild, as long as they are protected from pests and competitors. In other words, a plant in a garden comes closer to filling its fundamental niche.

From the idea of the niche, and the competitive exclusion principle, all sorts of things follow (including – I promise – some that have a direct bearing on invasions). First is the notion that, while three or four bakers in the same high street (or lizards in the same canopy) may be too many, one may not be enough, perhaps leaving some of the available resources unexploited. Following directly from this is the prediction that, all things being equal, communities with more species will tend to exploit the available resources more fully (and thus be more productive) than communities with fewer species – although the evidence for this theory is not as solid as generally assumed (as I explored in a previous book). Indeed, in a 2006 study Bradley Cardinale and others pulled together all the evidence available to show beyond reasonable doubt that monocultures (of plants or animals) routinely outperform mixtures of species:

> A fundamental tenet of biodiversity theory is that species must use resources in different ways to coexist stably. When species do coexist by such niche differentiation, theory predicts that diverse polycultures will produce more biomass and capture a greater fraction of limited resources than even the 'best' species monoculture . The balance of evidence from experiments does not seem to support this, and understanding why there is a divergence between empirical and theoretical conclusions is one of the foremost challenges in this field.

The first two sentences here are a succinct statement of niche theory and the effect this *must* logically have on the relative performance of mixtures and monocultures. The last sentence expresses the thinly veiled belief that the theory is correct, after all, and that surely it can only be a matter of time before we figure out why the evidence refuses to play ball.

TESTING NICHE THEORY

Another core belief that follows directly from niche theory is that niches are central to understanding why communities contain the number of species they do: specifically, that the available niches are all occupied and that the community is therefore *saturated*. In a saturated community, it's difficult by definition for a new species to gain a foothold. But if one does become established, that means fewer resources for the ones already present, leading sooner or later to the local extinction of a resident species. The almost universal finding of experiments, however, is that communities are *not* saturated and that local diversity can easily be increased. There are so many examples to choose from that one hardly knows where to start, but in one classic American study, sowing seeds of additional plant species increased the diversity of grassland plots from around 15 plant species to well over 20, and this difference persisted and showed no sign of declining for at least eight years. Nor is initially high diversity any particular barrier. In the extremely species-rich *alvar* grasslands of Estonia, even though tiny 10 x 10 cm plots initially contained 12 or 13 species, sowing extra seeds allowed three or four new species to be added to this total.

A slightly more subtle prediction of niche theory is that communities should exhibit *compensatory dynamics*. That is, in a community of finite resources, presumably containing a (relatively) fixed number of individuals, if one species increases its abundance, everything else must decline to compensate. Just to make sure we understand each other, note that predator and prey species will show compensatory dynamics for reasons that owe nothing to niches: more owls in a woodland will lead pretty quickly to fewer voles, and more lions to fewer zebras. We are talking here specifically about competing species on the same trophic level, such as our hypothetical community of tropical lizards. This is quite an easy idea to test, since plenty of people have monitored the numbers of individuals in communities of trees, fish, small mammals, butterflies or whatever for longer or shorter periods.

A recent study examined 41 such datasets and found an overwhelming preponderance of *positive* associations between species over time, i.e. the exact opposite of the pattern predicted by niche theory. As the authors conclude, this result doesn't rule out the kind of compensatory, negative interactions between species predicted by niche theory, but it does mean that such forces must be very weak compared to the effects of other variables such as weather that tend to increase or decrease populations of all species in unison.

NICHES AND INVASIONS

I promised earlier that all this would return us to invasions eventually, and it has. A corollary of compensatory dynamics, community saturation and more complete use of resources in species-rich communities is that such communities should be more resistant to invasion (less *invasible*) than species-poor communities. Over the last twenty years, an awful amount of time and effort has gone into testing this prediction, and the current view is ... probably not. Support for high diversity conferring low invasibility comes mainly from artificially constructed plant communities at small scales, but this kind of evidence begs all kinds of questions.

It's an open question whether the behaviour of randomly constructed experimental communities tells us much about the behaviour of real communities. Nor is it too surprising that the best evidence comes from small plots, since most communities start to look saturated if we zoom down to the scale of only a few individual plants. Evidence from the real world is nearly all in the opposite direction, with more-diverse communities actually being more likely to be invaded. Logically this is perhaps inevitable if we stop to consider the internal dynamics of communities. All communities, plant or animal, consist of many individuals, none of whom live forever, so there is inevitably lots of turnover. If new individuals are to establish, and thus the species persist in the community, the community must be 'invasible' by these individuals. Since as far as we know there is nothing special about 'invaders' (more about this in the next chapter), what applies to species already present must also apply to new species. In other words, a community must have been invasible to become diverse in the first place, and if it's to stay that way it must continue to be invasible. It can only have been wishful thinking that prompted a group of prominent ecologists to assert in 2002 that 'diverse communities will probably require minimal maintenance and monitoring because they are generally effective at excluding undesirable invaders'.

Given all this, would we expect alien species to be a major threat to the survival of natives? Plenty of people think we would, and are happy to quote a major 1998 report that found alien species to be the second largest threat to the survival of natives in the USA. Unfortunately the report is rather low on hard data (it's relatively easy to notice that a species is declining, much harder to determine *why* it's declining) and is strikingly biased by the inclusion of Hawaii. Exclude Hawaii and the USA is no different from Canada, where alien species are the *least* important source of threat, after habitat loss, pollution, overexploitation and other problems. A further problem is that most threatened species face more than one kind of threat, and it's very hard to tell which is the most important. Are exotic species really the main cause of decline, or are they just filling in the gaps left by pollution, climate change and habitat degradation?

But one thing is for sure: as of the end of 2007, there was no evidence that any plant had been lost from the USA, or from any US state, as a result of competition from an alien plant. In fact after 400 years of European settlement, establishment of aliens has led to an increase of about 18 per cent in the number of species in the US flora (excluding Hawaii and Alaska), while only 0.6 per cent of native plant species have been lost. In other words, a ratio of colonisation to extinction of about 24 to 1 for the USA as a whole, with similar values for individual states.

Even in Hawaii, much more heavily invaded, the ratio is 12 to 1. At the much smaller scale of US counties, the pattern is the same, with many new colonists and very few losses of native species. Moreover, the counties with the most natives always have been, and continue to be, the most invaded by aliens.

A survey of oceanic (i.e. remote) islands found that, as far back as records exist, they have been accumulating alien plants. In 1860 the average oceanic island had less than 1 introduced plant for every 10 natives. By 1940 the ratio was 1 alien for every 2 natives, and today the ratio is about 1:1. Despite all these new arrivals there have been very few extinctions among the original inhabitants, so the number of plant species on such islands has approximately doubled. Thus, although left to themselves remote islands tend to have rather few species (compared to similar continental areas at the same latitude), so many species have been introduced to Hawaii that it now has as many plants as a similar area of Mexico. Moreover, the evidence suggests that remote islands are by no means 'full' of plants, and that there is room for even more alien plants to establish, and thus for total plant diversity to increase: at the current rate the average oceanic island will have 3 aliens for every 2 natives by 2060. Do we have any idea how many different plant species might eventually be able to coexist on an island like Hawaii? No, we don't. Or, to express that conclusion in a more general form, in a report from US ecologists Dov Sax and Steve Gaines: 'we have a relatively poor understanding of the processes that ultimately limit how many species can inhabit any given place or area'.

Alien plants, generally speaking, seem to be little threat to natives. Nor, apparently, are freshwater fish; the richness of such species on Hawaii, for example, has increased by 800 per cent. Sometimes we simply have no idea if introduced species are a problem or not. Hawaii has at least 2,500 species of naturalised alien insects, but we don't know if this has caused the extinction of many native insects, or few, or none. It seems likely that fresh-water fish and (to a rather lesser extent) plants have had some difficulty in reaching remote islands, so maybe it's not surprising that the diversity of both on islands is lower than it 'should' be.

For birds, which are generally much better dispersed, the pattern is quite different, with colonists of oceanic islands approximately matching extinctions, so the total number of species present has not changed. It's tempting to see this as an example (at last) of a 'saturated' community, with natives driven to extinction on a one-to-one basis by competition from naturalised species, but in fact this is very unlikely. We know for certain that many native birds became extinct on these islands before the establishment of naturalised birds, and we also know that the primary causes of island bird extinctions are habitat loss and predation by introduced predators such as rats and cats. The story seems to be the familiar one of alien species taking advantage of a vacuum created by other aspects of human activity, rather than replacement of residents by colonists as a result of direct competition.

There's no evidence that things are any different in the sea; the 80 alien marine species introduced to the North Sea in recent centuries have caused no native extinctions so far, ditto for at least 70 species established in the Baltic, and the massive influx of species into the Mediterranean Sea via the Suez Canal has so far failed to cause more than a tiny number of local extinctions.

If most places seem able to provide space for more species than currently live there, then the introduction to a particular place of species that didn't formerly occur there (whether these are geographically 'aliens' or 'natives' is irrelevant) should generally increase local biodiversity, as countless experiments show it does. This may seem surprising, given the generally accepted negative impact of aliens, but it is exactly what is predicted by one of the most venerable ecological theories of all: the *theory of island biogeography* (see below). The theory, as its name suggests, was developed to explain the number of species on islands, but we now know it applies everywhere. Moving species around, so that there are more species in any particular place than there used to be, has exactly the same effect as moving an island nearer to the mainland, i.e. it increases the number of species that are *around* (in ecological jargon, increases the size of the source pool) and therefore capable of occupying any chosen location. Naturally, just as on real islands, this increased local biodiversity eventually increases the local extinction rate, but the result is a new equilibrium at a higher level of biodiversity than before.

The theory of island biogeography

Imagine a new, uninhabited volcanic island. It will soon start to acquire colonists, and the bigger the island, and the nearer the mainland, the faster this will happen. But as the number of species on the island increases, the rate of immigration of new species decreases, because any individual reaching the island is less likely to belong to a species that isn't there yet. Early on in the process of colonisation, the local extinction rate is low, because there's lots of empty space and even the most poorly adapted colonist can find somewhere to live. But as the island fills up, the local extinction rate increases as competition from the species already there increases.

The main prediction of the theory is that eventually an equilibrium is reached, where immigration equals extinction. This equilibrium depends on the island's size and distance from the mainland, which is why small, remote islands have rather few species, and therefore why importing a few more (effectively, the equivalent of moving the island nearer to the mainland) can easily increase local biodiversity.

ALIENS AND GLOBAL BIODIVERSITY

So much for local biodiversity, but what, ultimately, does all this mean for global biodiversity? As we move more and more species around the globe, accidentally or deliberately, the Earth gets more and more like its state 300 million years ago, when all the land was lumped together in a single supercontinent called Pangaea. We're getting more and more like Pangaea because shoving all the land together and moving species around are just two different ways of removing the barriers to dispersal that normally prevent kangaroos from meeting camels.

Again, the theory of island biogeography can help us to decide what this means for biodiversity, although care is needed to get the right answer. You can easily get the wrong answer by counting the cumulative number of plant species as you look at larger and larger fractions of the British Isles (or bird species in North America, or whatever), and then extrapolating to an area of land equivalent to the whole globe. The result is many fewer plant species (or birds) than the Earth currently supports. Why is this wrong? It's wrong because different bits of Britain (or North America) have overlapping floras and faunas (the extra area has species in it that you counted already), meaning that as you look at larger areas you find lots of species you counted already, so you add new species at a relatively slow rate. But recreating Pangaea involves adding together chunks of land that have nothing in common (the floras and faunas of North America, India and Australia are completely different), and the actual rather pleasing result of doing that is … nothing.

That is, the biodiversity of Pangaea exactly equals the sum of the present biodiversity of the Earth's various continents. It may seem surprising that homogenisation does not necessarily mean low diversity, but we don't need to travel back in time to Pangaea to have that confirmed. Because ocean currents easily transport fish and coral larvae over distances of thousands of kilometres, coral reefs over vast areas are effectively closely connected and contain mostly the same species, yet remain one of the most diverse communities in the world. Both theory and coral reefs agree: 'In the future, different regions of the world will be more similar than they are now. They will also be more diverse', concluded Mark Davis in *Bioscience*.

However, before you break open the Champagne at the news that moving species around at our present rate will increase local biodiversity, and have little or no effect on global biodiversity, it's worth recalling that that's not all we're doing. We have already altered three-quarters of the Earth's ice-free land surface, and by 'altered' I mean made it much less able to support the animals and plants that used to live there. Again island biogeography tells us what that means: if we leave a quarter of the Earth for wild animals and plants, we eventually end up with a quarter as many species as we started with. Many of these survivors will be species that we have moved to new continents, and many of them will be species that don't too much mind a planet that consists mostly of crops, cities and roads. But blaming them for that sorry state of affairs would be foolish.

The kauri story

Kauri (*Agathis australis*) is an iconic New Zealand conifer, long famed for its size, longevity, useful resin and fine timber. Old, mature trees have remarkably stout, more or less cylindrical trunks that branch only when they reach the canopy. Much kauri forest has been logged, but kauris are generally found as mixed populations of a few big old trees, scattered younger trees, sapling and occasionally dense patches of seedlings. Since kauris of all ages are markedly intolerant of shade, and thus incapable

New Zealand's iconic kauri tree.

of regenerating from seed under a dense tree canopy, it was for many years a mystery how these mixed populations arose.

The story was eventually figured out by ecologist John Ogden. Dense populations of kauris establish after major distur-bances such as fires, storms, landslips or vulcanism, when there is plenty of light for the young trees. These dense populations cast deep shade and thus prevent any more kauris establishing from seed for a long time – there may be frequent crops of seedlings, but these soon die in the shade of the forest floor.

Kauris live for centuries, and old trees may be well over 1,000 years old, but old trees eventually senesce and begin to die, creating gaps in the canopy over a period of a century or two. Many of these gaps are occupied by more-shade-tolerant broadleaved trees, or by other conifers, but enough are captured by kauris to establish a second cohort (albeit smaller than the original and rather ragged). This second cohort goes through a similar process, eventually being succeeded by an even smaller and less synchronous cohort of young kauris.

Three cohorts of trees, each one smaller than the last, is about the limit, and if there is no major disturbance for a millennium or more, kauri eventually goes extinct locally. But another catastrophe can reset the whole process at any point, and since in practice there's always a fire or a disastrous storm in the 1,000 to 2,000 years it takes three cohorts of kauri trees to expire, kauri forest persists indefinitely.

LESSONS FROM HISTORY

Before we move on, here's one more interesting question: why has the spread of alien species revealed the established niche-based view of the world to be so wide of the mark? There seem to be two problems with that traditional model.

First, the number of species in any one place seems to be limited more by the number that can get there, i.e. by dispersal, than by the number that could live there if dispersal were no problem. I'll return to this topic in the next chapter.

Second, few ecological communities remain undisturbed for long enough for niche-based processes to arrive at any kind of conclusion; essentially most communities, however permanent they may appear, are in reality still recovering from the last disturbance. The kauri example above may be an extreme case, but it illustrates well that figuring out what is going on from a 'snapshot' of any plant community at a single moment is frequently impossible, and thinking you can make sense of a community in terms of niches being neatly parcelled out like the ones in your ecology textbook is wrong a lot more often than it's right. You cannot hope to understand the relationship between kauri trees of different ages, or between kauri and other tree species, unless you think about how the whole forest started out, maybe 1,000 years ago, and what has happened to it since.

Sometimes a dispersal problem is linked to recovery from a remote historical event. For example, several recent studies have noted an increase during the twentieth century in the numbers of plant species found on the high mountain summits of the European Alps, an increase more or less explicitly attributed to upward migration in response to recent climate warming. But nearly all of these species already occur at the same (or higher) altitudes on other mountains. A better explanation is that Alpine species are slowly catching up with the increase in temperature following the Little Ice Age, a period of intense cold that lasted from about 1550 to 1850 (the first 'frost fair', held on the frozen Thames in London, was in 1607, the last in 1814). Extreme low temperatures and continuous snow cover depressed the upper limits of Alpine plants, and since most aren't very well dispersed, they still haven't got back to where they were before the Little Ice Age. Modern anthropogenic climate change gets the blame for nearly everything climate-related these days, but the obvious answer isn't always the right one.

CHAPTER SEVEN

SPOTTING THE BAD GUYS

Most species transported to new countries or regions fail to establish at all, and of the minority that do only a very few go on to become really abundant and eventually have any environmental or economic impact. Not surprisingly therefore, the Holy Grail of invasion biology has long been a reliable way of deciding, in advance, which species are likely to go on to cause problems. If a way of doing this could be found, it holds out the prospect of deciding which species should be allowed into a country and, among those already present, distinguishing those that require prompt action from those that can safely be left to themselves.

Ideally, the search for the important characteristics of good invaders would involve comparing successful invaders with unsuccessful ones. Unfortunately, we only ever discover the identity of those that go on to become successful; by definition, unsuccessful accidental invaders simply vanish without ever being noticed. There are two ways round this. One is to compare successful invaders with species from the invader's home region that have not invaded the particular region of interest. The obvious drawback here is that the stay-at-homes may never have been introduced outside their home region, and thus their invasive abilities remain untested. Another option is to compare the successful invaders with natives from the invaded region, and in practice this is the option that has been overwhelmingly favoured by researchers.

WINNERS AND LOSERS

My own original contributions to the science of invasions have been few and generally not of any great importance. But I have to report that, surveying in the early 1990s – at more or less the beginning of the attempt to find out what's special about invasive species – it occurred to me that people were tending to ask the wrong question. I had already shown (although this was far from an original observation) that, among native British plants, there were very big differences between species currently expanding in range and abundance ('winners') and declining species ('losers'). These differences were consistent with the overwhelming impact of modern intensive agriculture and were particularly obvious in England (as opposed to Wales and Scotland), where those impacts are greatest. Winners were big, fast-growing, fecund, early-maturing plants, well able to cope with the modern fertile, disturbed agricultural and urban landscape. Losers were the exact opposite: often small, slow-growing, long-lived, poorly dispersed and essentially confined to declining fragments of an older, less-intensive agricultural landscape.

So far, so not very surprising, but the implications for invasions are profound. Successful invaders are species that, from a standing start, become common enough to be noticed in a short space of time and are therefore by definition winners. Natives, on the other hand, are a complete mixture of winners and losers, plus plenty of others that aren't noticeably either. Thus a straight invaders/natives comparison has an inevitable bias built in at the outset: compare a bunch of winners (invaders) with another collection of species that, while mixed, contains a significant proportion of losers (natives), and you will find that the two groups are different, with the invaders, on average, enriched in the well-known characteristics of winners.

Just in case this wasn't obvious enough, I published a paper in which I compared invaders with successful, *increasing* natives. In almost all respects they were the same, and the few in which they weren't could be attributed to climate, reflecting the rather trivial fact that most invaders in Britain come from warmer climates. Thus, I concluded: 'These results are consistent with the view that in most key respects increasing alien and native species are functionally similar, and that the difference between local and long-distance range expansion is merely one of degree. This similarity between aliens and natives contrasts with relatively large functional differences between decreasing species (virtually all natives) and increasing species (a mixture of aliens and natives) in England and the Netherlands.'

Given that I wrote this in 1995, and that the paper in which I said it has been reasonably well cited (and maybe even widely read, although you can never be sure), it's a source of continuing astonishment that literally hundreds of papers since then have made the same erroneous invasive/native comparison, and come to the same inevitable and distinctly un-illuminating conclusion as the very first comparisons. That is, they've often found not very much, depending on the details (invasive aliens are always winners, but sometimes by chance the natives are mostly winners too, and in that case you don't find much difference), but when they *do* find a difference, it's exactly the same as the difference between winners and losers that I (and others) pointed out decades ago.

By the way, in case you're wondering, am I claiming to be the only person on the planet who actually understands what's going on? No, I'm not. Over the last few decades a scattering of other researchers have had the same bright idea of comparing invasive aliens with successful, common, expanding natives. And what have they found? Usually, nothing. Successful species certainly differ from unsuccessful ones, but being native or alien has nothing to do with it.

At this point, a rather philosophical question may or may not have occurred to you, but I'll deal with it anyway. Why do those who compare invasive and native species keep finding (if they find anything) the same differences? Would we expect to find a consistent difference between winners and losers, across the globe? Maybe not, and to explain why not, consider the analogy of species' temperature preferences. If we looked at the temperature preferences of millions of species across the whole planet, would we expect to find that successful, expanding species had a consistent preference for warmer or cooler temperatures? Perhaps neither; in fact, as long as we had a representative sample of species from the equator to the poles, we would probably find, on average, a liking for something close to the average temperature of the Earth. Only if the temperature of the Earth were changing might we expect something different – if the Earth was warming, then we might expect expanding species to share a preference for warmer temperatures. In a similar way, we find a consistent picture of the typical 'winner' because the Earth *is* changing, in fact has changed, with more than 75 per cent of the Earth's ice-free land now altered by human land use.

The view of the biosphere you learned in school, of natural biomes, disturbed to varying extents by humans, is seriously out of date. The modern world is essentially a mosaic of new 'anthropogenic biomes' (croplands, plantations, settlements, cities, rangelands), with here and there natural ecosystems embedded within them. It's because so much of the world has been transformed so dramatically that there is now a consistent 'winning' syndrome. In a world before significant human influence, there was room for everyone, and all available tickets in life's lottery had some chance of winning. It's only in the last few thousand years, and especially in recent centuries and decades, that being a rat or a weed has turned out to be *the* golden ticket in life.

To reiterate, and to make sure we don't lose sight of the basic point, winning and losing has little or nothing to do with being alien or native. Invasive aliens are winners by definition, but so are many natives. If, as in the UK, we profoundly transform the landscape in a way that suits a small suite of fast-growing, effectively dispersed plants, then that's what we get, with the aliens that fit this description expanding along with the natives. If Britain had no alien plants, we would still have a landscape dominated by plants that no one likes very much, except they would be exclusively natives: coarse grasses, nettles, brambles, ragwort, docks, dandelions, thistles, bracken and hogweed. In the UK, as elsewhere, successful species, alien or native, are symptoms of change rather than drivers of that change, and all they are telling us is that they are very pleased with the changes we have made to their environment. We have the plants (and animals) we deserve.

TWO RATHER UNSUCCESSFUL THEORIES

Despite – or maybe even because of – the activities of those who cannot resist adding to the ever-growing pile of data that shows that species (native or alien) that like a world mucked up by humanity are doing just fine, and that those that don't are going down the plughole, there are other theories out there that attempt to explain invasiveness. These focus on the invasion process itself rather than on the traits of invaders, and the front runner is the *Enemy Release Hypothesis*, or ERH for short. The ERH assumes that invasive species do better in their new habitat because they have left some or all of their specialist natural enemies behind, whether predators, pathogens or parasites. Studies that simply count the number of different natural enemies in native and introduced ranges routinely find support for the ERH (i.e. fewer enemies in the introduced range), but almost all suffer from a sampling problem. Natural enemies are nearly always better documented in the native range, and in any case it's the enemies found in the source population that's important, not in the whole range. Once this is accounted for, most support for the ERH disappears. For example, one study found fewer parasitic worms in introduced starlings in North America than in the entire native range of Europe and Asia. But once allowance was made for the actual local source of the starlings, the difference disappears: various evidence suggests starlings arrived in North America via Liverpool, and American starlings have most of the parasites of Liverpool starlings, plus quite a few others, either American natives or European parasites introduced with other birds. In fact, American starlings have *more* parasites than are found in the likely source population.

In any case, whatever the studies that simply count natural enemies find, those who have looked for actual increases in growth or survival of introduced species mostly find little support for the ERH. One reason for this is probably the activities of native enemies of introduced species. The ERH depends on a general unwillingness or inability of native predators or parasites to attack introduced species. Yet the same people who believe in the hypothesis are just as likely to assert that such predators or parasites would wreak havoc on 'naïve' prey if *they* were introduced to a new habitat. These two ideas can't both be right. Some of the best evidence comes from studies that compare introduced and native species with and without the available natural enemies in the invaded habitat. Without natural enemies, both perform about the same, and although natural enemies have a negative impact on both they have a bigger impact on the aliens, i.e. the opposite of the prediction of the ERH.

Some particularly unlucky invaders even run into other non-native predators that arrived before they did. A nice example is the introduced cricket *Teleogryllus oceanicus* in Hawaii. The cricket is attacked by a non-native parasitic fly that locates male crickets by their mating calls. Since parasitised crickets eventually die, the intense selection pressure has led to the evolution of silent males, which can now only mate by lurking next to one of the few surviving singing males and intercepting females attracted by him. Once the last singing male is found and killed, it looks like curtains for the cricket.

A variation on the ERH is the *Evolution of Increased Competitive Ability Hypothesis* (EICAH). The ERH is supposed to provide instant benefit to invaders, while the EICAH proposes that if invaders are freed from the attentions of their specialist natural enemies, they can evolve to use any resources saved from the need to defend themselves to grow bigger and faster. Given the general lack of support for the ERH, you'll not be surprised to learn that EICAH doesn't seem to work very well, either. The hypothesis predicts, for example, that plants introduced to new countries would at least manage to grow taller than they did at home. Yet one classic study of European plants in the USA, and American plants in Europe, found little difference in size, and the few differences there were tended to be in the 'wrong' direction, i.e. plants were taller at home.

A SLIGHTLY BETTER THEORY

The net result of all this feverish activity is a dawning realisation that there's nothing very special about invaders and nothing very special about the invasion process. So if some species really do seem to be much better invaders than others, why is that? Well, if you recall from the last chapter, one reason for the failure of niches to explain the diversity and composition of ecological communities is dispersal, or rather its failure. Many species aren't found in places that look (and indeed are) quite suitable because they just never got there. But one of the features of the modern world is our propensity to move plants and animals around the planet, sometimes deliberately, often accidentally, thus breaking down all the normal barriers to dispersal. Which brings us to yet another invasion theory: *propagule pressure*, which suggests that the best invaders are simply those that happen to have been introduced most often and in the largest numbers.

Given the potential of propagule pressure to make all other theories redundant, it is surprising how rarely it's been considered as an explanation for differences in invasion success. One recent review looked at 79 separate studies that attempted to find the characteristics of successful invaders, with the usual abysmal results. Of these studies, only 13 even considered the role of propagule pressure. Because two of these 13 studies looked at more than one species, they actually considered 15 species, and in 13 of these (in other words, in almost all of them) propagule pressure was found to be important. It did not matter whether plants, birds, fish, mammals or invertebrates were studied – there was strong support across the board for the idea that higher propagule pressure means greater invasion success.

Once you start looking for it, evidence for the importance of propagule pressure is everywhere. Nature reserves with more visitors are more invaded than those with fewer, because visitors are the primary source of (mostly unintended) introductions. Attempts at biological control, which is essentially assisted invasion, routinely report that introducing lots of individuals is vital for success. One could hardly wish for a better example of the power of propagule pressure than *Rhododendron ponticum* in the UK (see below). Contrast the behaviour of *R. ponticum* in Britain and Norway. With its acid soils and mild, wet climate, southwest Norway is rhododendron heaven, and numerous species and cultivars (including *R. ponticum*) are popular garden plants there. But *R. ponticum* never had the massive planting in the wild that it had in Britain, and it's not invasive in Norway.

Rhododendron ponticum: Special clearance offer, 105 shillings per 1,000

Rhododendron ponticum was introduced to Britain in 1763. It is a native of the area around the Black Sea and also Spain, from where British plants originate (and where it is now rare and indeed endangered). It's normal among those who spend their time plotting its downfall to attribute the success of *R. ponticum* in Britain to its biology: ericaceous mycorrhizas that enable it to grow well on poor, acid soils; toxic leaves; abundant small, wind-dispersed seeds; ability to resprout readily when cut back, and from underground buds if the above-ground parts are killed by fire. Yet this is more or less a description of the genus *Rhododendron*; none of these things is unique to *R. ponticum*. You might argue that the British plant is a hybrid, but then so are thousands of cultivars, many of them with *ponticum* in their parentage; there's nothing intrinsically invasive about hybrids.

Why are none of the 400 or 500 other species of *Rhododen-dron*, and thousands of hybrids and cultivars, grown in Britain a problem? Only one other species, *R. luteum*, is even naturalised, and is not a conservation problem. In other words, the search for the traits that explain invasiveness is once more in vain, as it nearly always is. Could the answer be propagule pressure? All the evidence suggests it is.

Rhododendron ponticum sets abundant viable seed and is easy to propagate. It was also just plain lucky, being available at exactly the right moment to take advantage of the British fashion for wild and woodland gardens promoted by William Robinson, and to provide cover for game on shooting estates. Shooting game became increasingly popular throughout the nineteenth century, and woodlands became more and more to be seen as pheasant coverts rather than sources of timber; gamekeepers took over from foresters. Trend-setting gardener J. C. Loudon wrote in 1838: 'In Britain, it is planted as an ornamental shrub, not only in open situations, but, on a large scale, in woods, to serve as undergrowth, and as a shelter for the game'.

As an article in *The Gardeners' Chronicle* in 1841 commented: 'It is very easy to fill woods with them, by sowing the seeds broad-cast ... A man and boy can collect enough [seeds] to sow acres in a few hours.' *Rhododendron ponticum* was the perfect candidate for both purposes, since its ease of propagation meant the nursery trade could supply it in large quantities at low prices, or landowners could do it themselves. Cheapness and abundance also made *R. ponticum* the species of choice for grafting choicer varieties, and hundreds of thousands of plants were used every year for this purpose. Not that it was actually much use, since it tended to sucker and overwhelm the more delicate scion. Much *R. ponticum* originated in this way.

In short, *R. ponticum* was given the mother of all oppor-tunities to make itself at home in Britain, and the fact that it

Rhododendron ponticum thriving in Britain.

did exactly that (and that other species didn't) should surprise
no one. As Katharina Dehnen-Schmutz and Mark Williamson
put it: 'It was selected and hybridised for hardiness. Its spread
and increase was from propagation by nurseries and estates. It
was distributed over distances far greater than its seeds could
travel naturally. It was brought directly to habitats offering the
most suitable conditions for its survival. Without all this the
plant might perhaps still exist in the British Isles today just
as specimens in botanical and horticultural collections like
thousands of other introduced plants.'

As usual, we have no one to blame but ourselves.

ACCLIMATISATION SOCIETIES

Some of the best evidence for propagule pressure comes from deliberate introductions by acclimatisation societies, which were particularly active in European colonies in the Americas and Australasia. Such societies were keen to introduce familiar species, partly to make their new country seem more like home, and partly from a misplaced belief that the local flora or fauna was in some way deficient or incomplete. In retrospect, and given the havoc subsequently caused by some of the deliberate introductions, such as the rabbit in Australia, it's perhaps fortunate that lack of money and biological knowledge meant that many introductions did not succeed.

New Zealand acclimatisation societies were particularly keen to introduce European songbirds, and there were many regional societies that kept detailed records of exactly what they introduced, and with what success. The results are clear: if 10 or fewer birds were introduced, failure was almost guaranteed, while introductions of more than 100 birds were almost always successful. Between those two extremes, introducing 11 to 100 birds led to successful establishment around half the time. Notice that it scarcely matters whether we are talking about skylarks or linnets, house sparrows or cirl buntings – differences between species are merely notes in the margin compared to the overwhelming effect of numbers of birds introduced.

So, at the risk of repeating what must by now be a familiar refrain, it seems we can add one more to the already long list of things we don't understand about why some species succeed and others fail, how species fit together into communities, and how those communities behave.

CHAPTER EIGHT

OUT OF CONTROL

Attempts to control species suspected, rightly or wrongly, of causing economic or environmental harm have a long and far from illustrious history. Things have rarely turned out quite as planned, and the cure has often turned out to be worse than the disease. Those described in this chapter will give you a flavour of the problem, but one could document many, many other examples. You'll find, by the way, that much of the action takes place on islands, for one simple reason. If eradication of an introduced species is the aim, as it often is, then this is much easier if the species in question is confined to a small area. Thus small islands are one of the few places where eradication is a realistic ambition; once successful aliens are loose on large land masses, they are usually there to stay, like it or not.

Of course, perhaps a large part of the problem is that, as we've seen in the previous chapter, we are so poor at figuring out why some invaders are more successful than others. Nor in accepting that much of the difference may lie in *our* behaviour. Which, of course, is hardly the best platform from which to embark on the attempt to control invasive alien species.

ALIENS AND ISLANDS

It can often look like an introduced herbivore is damaging native plants – but then removing the herbivore often reveals a more complicated story. For example, when pigs and goats were removed from Sarigan Island in the Northern Mariana Islands, an alien vine, previously the goats' favourite food, quickly took over most of the island. Similarly, when feral cattle were removed from San Cristobal Island in the Galapagos, the exotic guava – previously suppressed by grazing – grew rapidly into dense, impenetrable thickets. It was much the same story when cattle were removed from Californian grasslands in an attempt to protect native wildflowers. The principal beneficiaries were alien grasses, which reduced the abundance of native wildflowers to below the level before the cattle were removed. In New Zealand, livestock were removed in an attempt to restore habitat for the threatened Whitaker's skink (*Cyclodina whitakeri*) in its last mainland site. But subsequent monitoring revealed that the skink continued to decline, for reasons that turned out to be far from simple. Reduced grazing favoured dense growth of introduced pasture grasses, in turn resulting in periodic outbreaks of rodents, leading to increased numbers of predators, which then ate the skinks.

More mistaken introductions, followed by equally mistaken attempts at control, have probably occurred on Hawaii than anywhere else on Earth. As a result of concern for the survival of subalpine tropical dry woodland dominated by two endemic tree species, an area of over 4 sq km was fenced in 2001, followed in 2003 by removal of exotic feral goats from inside the fenced area. This was a large and expensive operation and by 2009 the results were clear: vegetation cover increased in the area cleared of goats, but all of the increase was accounted for by exotic species. Native species either didn't change or actually declined. The main beneficiary was *Senecio madagascariensis*, an exotic annual that increased from barely detectable to almost 15 per cent of total vegetation cover. None of this happened in a control area where goats still grazed. *Senecio madagascariensis* produces alkaloids that are toxic to most herbivores, but do not deter goats. Indeed goats have been used for (effective) control of *S. madagascariensis* in Australia, so although the outcome of removing them in Hawaii was depressing it shouldn't really have been a surprise.

Often the negative impacts of removing an alien species involve complicated (and unforeseen) feedbacks. Removal of rabbits from Macquarie Island in the Southern Ocean led to a big increase in the cover of a native tussock grass, which also happens to be the favourite habitat of the introduced black (or ship) rat, increasing the threat of rat predation on nesting birds. In Hawaii, although removing feral pigs and sheep allowed some native plants to recover a little, again there was a big expansion of alien grasses. Because these grasses are highly inflammable, this increase led to more fires, consequent loss of native woodlands, yet more fires, and so on. Removal of sheep and cattle from Santa Cruz Island off the coast of California led to an explosion of fennel, thistles and other introduced plants. These in turn allowed a big expansion of feral introduced honeybees, making their eventual eradication more difficult.

A common feature of many of these expensive failures is an attempt to tackle exotic species one at a time in ecosystems that have been dramatically altered by numerous introductions. In New Zealand, introduced rats and possums form a large part of the diet of stoats, another introduction. When efforts were made to remove all three by poisoning the prey species, the stoats, in the absence of previously abundant exotic prey, turned their attention to native birds and their eggs.

Introduced feral cats occur on many islands, often coexisting with introduced prey such as rabbits, rats and mice. The cats usually have a big impact on the populations of these exotic prey species, strongly preferring rabbits if they're present, and rats if they're not. Trying to remove the cats without simultaneously removing their prey (or vice versa) nearly always has unpleasant (but not exactly unexpected) consequences. Before their removal from Marion Island in the Indian Ocean, cats were the main predator of exotic house mice, which in turn ate a flightless endemic moth, which rather surprisingly turned out to be largely responsible for nutrient cycling on the island. Following the eradication of cats, the long-term consequences of the increase in mice and the decline of the moth are unknown.

The case of eagles and pigs on Santa Cruz Island, in the California Channel Islands, reveals the crucial importance of removing invasive species in the right order. The critically endangered island fox (*Urocyon littoralis*) declined precipitously in the mid-1990s, for reasons that were not in doubt: predation by golden eagles. Golden eagles are of course a native North American species, but they were attracted to the island in large numbers by the presence of non-native feral pigs. The answer was clearly to remove the pigs, but ecologists predicted that, if pigs were removed first, the large eagle population would focus more on the foxes, probably leading to their extinction. No one wanted to harm the eagles, so a programme of live trapping was begun, resulting in the removal of 44 eagles from the island; only then was it considered safe to eradicate the pigs. Nevertheless, the foxes hardly recovered, and it was

Santa Catalina's endangered island fox.

subsequently discovered that not only had one pair of golden eagles been missed, this pair happened to be fox specialists. Once this pair of eagles was tracked down and removed, there was a modest recovery of the foxes, although numbers remain low. Underlying the whole saga, as so often, was a man-made environmental problem – in this case, the loss of bald eagles from the island during the 1950s owing to contamination of the surrounding sea with DDT. The highly territorial bald eagle competes with the golden eagles for nest sites, so it was not only the abundant pigs, but also the absence of competitors, that allowed golden eagles to move to the island in large numbers. Bald eagles have now been reintroduced to the island, and with any luck the foxes will now be safe. Bald eagles themselves take the occasional fox, but are primarily fish-eaters.

Although there may be interactions with introduced plants, all the above examples concern the removal from islands of introduced animals. For one very good reason: plants are not easy to eradicate. The Galapagos Islands illustrate the problem. The Galapagos archipelago, unlike many other islands, is still considered to be relatively pristine, with 95 per cent of its pre-human biodiversity remaining. Nearly all (97 per cent) of the archipelago is a National Park, and most introduced plants still have limited distributions, mostly around farms and gardens. Its status as a World Heritage Site and popular wildlife tourism destination also means that the political climate in Galapagos is favourable. In short, if eradication of introduced plants can be achieved anywhere, it can be achieved on the Galapagos. So it's not surprising that the Global Environment Fund (along with other partners) put together a six-year (2001–2007) $43 million programme entitled 'Control of Invasive Species in the Galapagos Archipelago'. The programme had several aims, but among them was the eradication of 23 introduced plants (30 individual projects), all of them known or suspected to be invasive, and all of them, crucially, with relatively limited distributions.

So, now that the dust has settled, how did it go? All in all, not very well. Five planned projects never began, since further investigations revealed the plants concerned were more widespread than thought, making eradication impractical. Most attempted projects also failed, the most common reason being lack of cooperation from landowners. Often this was because the target plant was considered to be useful, for example, for medicine or timber; after all, most of these plants were introduced for a reason. But there were other problems; several farmers revoked permission midway through the project, accusing the field workers of stealing poultry and wood from their farms. Sometimes a project failed because only one of several landowners failed to give permission. Other projects ran into biological problems, such as a persistent seed bank, or simply found that the plant was spreading faster than it could be found and removed. Often time and money simply ran out before eradication could be completed; the lesson from previous successful eradication programmes is that one or more decades is usually required (as with the devil's claw example below).

In the end only four out of thirty projects were successful, all where the conditions were unusually favourable: plants with very limited distributions, no persistent seed bank and land controlled by a single owner. None of this should have been surprising; a study of attempted plant eradications in California showed that plants that occupy an area of no more than one hectare can usually be eradicated, but the probability of success declines rapidly as the infested area increases beyond this. One hectare is a tiny area (only slightly larger than a football pitch), so in practice this means eradication has to begin long before there's any possibility of knowing whether a plant will eventually cause any kind of problem – in fact, almost before the plant has been detected.

Optimists have likened eradication to a 100 per cent lump sum payment, after which the job is done and no further payments are necessary. Experience, on the other hand, suggests most attempted eradications have more of the character of a small downpayment, with a promise of further instalments that continue indefinitely.

A MAINLAND EXAMPLE: THE DEVIL'S CLAW

We've seen how difficult it is to eradicate alien plants even from relatively small islands. On large landmasses, the problems are much greater, but nevertheless the idea of eradication remains attractive – if successful, it has the undoubted attraction of solving the problem once and for all. An interesting example is the attempt to eradicate the devil's claw (*Martynia annua*) from Gregory National Park, Northern Territory, Australia.

Devil's claw is a Mexican native and is widely grown around the world for its attractive flowers and the distinctive hooked fruits that give rise to its common name. It has now become naturalised in many warmer parts of the world, including several parts of Australia. In the late 1980s a programme began with the intention of eradicating devil's claw from the National Park. Over 20 years later the project remains the longest known continuously funded eradication project of this scale in Australia, and one of the longest-running in the world. It's certainly one of the best documented. Up to 2010 the programme had removed 107,910 individual plants and cost an estimated Aus $531,348 at 2009 prices (more if volunteer labour is costed).

Are we, after all that, any nearer eradicating devil's claw from the Park? The simple answer is no. In the light of the known facts about devil's claw, was eradication ever likely? Again, the answer is no. Devil's claw is an annual, which grows and sets seed quickly, its seeds persist in the soil for around five years; the area infested is huge, difficult to search and often inaccessible owing to flooding. On several occasions, parts of the infested area could not be searched because of flooding, allowing the plant to set seed and effectively setting the programme back to square one. Despite the huge cost of the eradication programme, success would have required much more investment – for example, in helicopters.

Note that in any case, it was only ever eradication from the *Park* that was envisaged; no one has even suggested that devil's claw could be eradicated from Australia. Given the plant's ability to disperse great distances attached to the coats of wild or feral mammals, reintroduction from outside any cleared area would always be possible, indeed probable. Nor is that the only way it gets around; it originally arrived in the Park in contaminated hay.

Finally, what was the cause of all this hoopla? All the Queensland Government's fact sheet has to say is that devil's claw is 'regarded as a significant environmental weed in the Northern Territory'. In practice it is, at the very worst, a minor annoyance to the cattle industry; there is no evidence whatsoever that it is able to establish dense stands, changes the character or nature of native ecosystems or is a threat to the survival of any native animal or plant. My opinion, for what it's worth, is that *Martynia* has the misfortune of being

The unfortunately but aptly named devil's claw, *Martynia annua*.

called devil's claw. If it had been called, like its compatriot *Hunnemannia fumariaefolia*, something pink and fluffy like the Mexican tulip poppy, we could have arrived exactly where we are now without spending Aus $500,000.

Australia, of course, does not have a monopoly on attempting to control species that probably don't deserve all the attention. In the USA, we've already noted the spleen provoked by tamarisk and purple loosestrife. But in the Intermountain West, the region between the Rocky Mountains on the east and the Cascades and Sierra Nevada on the west, two other species are slugging it out for the position of public enemy number one: spotted knapweed (*Centaurea stoebe*) and cheatgrass (*Bromus tectorum*). In terms of litres of herbicide used, number of hectares treated and hours of labour expended, spotted knapweed is probably just ahead. It certainly has a claim to be the most heavily managed alien plant on public land in the region, designated a noxious weed in 11 states. Yet a recent three-year study found the effect of knapweed on cover and richness of native plants to be negligible. On the other hand, the herbicide commonly used for its control is far from harmless, causing a significant loss of native wildflowers and, ironically, encouraging invasion by cheatgrass, a plant that may well be a real problem.

USEFUL ALIENS

Sometimes eradication efforts are halted by the discovery that the target is actually doing something useful. From the 1960s to the 1980s, the US Department of Agriculture (USDA) introduced many alien honeysuckle species in land reclamation projects, and to improve bird habitats. They then changed their mind, declared several species of introduced honeysuckles to be harmful and banned their sale in more than 25 states.

But they may have been right in the first place. Recent work in Pennsylvania has found that more non-native honeysuckles means more native bird species – three to four times as many native birds in one site. Also the seed dispersal of native berry-producing plants is higher in places where non-native honey-suckles are most abundant.

New Zealand is short of insect pollinators, so vertebrate pollination is common; even in species that look like they should be insect-pollinated, vertebrates often make an important contribution. A study of three New Zealand woody plants found that their flowers were visited by a bat, a gecko and five native bird species. All these pollinators survive on offshore islands, but are now rare or extinct on the mainland, where the flowers are now visited – and pollinated – by the ship rat and by an alien bird. So even though the ship rat was probably partly responsible for the loss of some of the native vertebrates it's now doing the job they used to do. Ship rats are unlovely and unloved, and that won't be enough to save them from attempts at extermination, but it's worth knowing that such attempts may have unintended consequences; even 'problem' aliens may turn out to have their uses.

It's also true that in a world increasingly altered by humans aliens may sometimes simply be the best species for the job. A nice example can be seen in the Don Valley Brick Works, the abandoned 40-acre site of Ontario's longest-functioning brickyard, which is now an important part of Toronto's network of green spaces. In its early years the site was managed by the Toronto and Region Conservation Authority (TRCA), and is currently co-managed by the City of Toronto's Parks and Recreation Department and Evergreen, a Canadian NGO that strives to conserve natural and cultural landscapes. The mission of all these bodies has been to restore a clean, green and accessible Don River watershed.

Creeping thistle – a welcome sight?

The TRCA quickly uncovered the buried course of Mud Creek (a tributary of the Don) and created a series of connected ponds with trails, bridges, and viewing platforms, transforming a rubble- and rubbish-strewn quarry into an attractive wetland. The only fly in the ointment was that the new park's woody vegetation was dominated by the alien Manitoba maple (*Acer negundo*, often known as box elder), and its herbaceous vegetation by European Canada thistle (*Cirsium arvense*, or creeping thistle). Volunteers piled enthusiastically into thistle control, toiling to hand-pull the tenacious, deep rhizomes of Canada thistle and plant a variety of native wildflowers on the brick-and-rubble slopes of the former quarry. About 80 per cent of the volunteer work was devoted to invasive-species management, focused primarily on pulling up and cutting thistles, and collecting and bagging thistle seed heads. This work was unpleasant for the volunteers, but they were happy to do it, until they returned to find Canada thistle regenerating everywhere and no trace of live wildflowers. Maybe the thistle was the best anchor for the thin layer of imported topsoil, the best (only?) option for vegetation of any sort on these unstable slopes? Both chastened and enlightened by their attempt to work against – rather than with – the grain of nature, the volunteers began to intermingle small patches of wildflowers within the Canada thistle, carefully supporting and encouraging them to gradually spread out across the slope. Before long the uncooperative rubble slope was a patchwork of thistles and expanding patches of self-sustaining native wildflowers.

The Brick Works example illustrates an increasingly common problem: what to do with a site that is so altered that the whole idea of aboriginal, native vegetation simply has no meaning? Under such circumstances, treating the successful aliens (which are not going away, however hard you try) as allies rather than foes may offer the only way forward.

Here is another example. In much of the temperate northern hemisphere, *the* stabiliser of mobile sand dunes is marram (*Ammophila arenaria* in Europe, *A. breviligulata* in eastern North America), and introduced *Ammophila* often replaces native dune species in other parts of the world. But sometimes the boot is on the other foot. *Carex kobomugi* (Asiatic sand sedge), native to coastal areas of Japan, China, Korea and Russia, was widely and enthusiastically planted for dune stabilisation along the east coast of North America from the 1960s through the mid-1980s. The sedge was initially promoted because it is more disease-resistant, more tolerant of trampling and will grow in a wider range of habitats than native beach grasses. Its spiky flowers and sharp rhizomes also form a natural way to deter foot traffic from leaving established

141

trails. All in all, it was seen as an ideal species for areas needing dune stabilisation.

Asiatic sand sedge did everything that was asked of it, and dunes stabilised by it have proven to be less damaged by storms than similar dunes stabilised by marram. Nevertheless, by the late 1980s it had fallen foul of the general reaction against 'too-successful' alien species, planting ceased by the early 1990s, and by 2000 it had plummeted to being one of the '10 most unwanted plant species in New Jersey'. Note that, as usual, all this happened without any evidence (or even any attempt to obtain any evidence) that the plant was causing any harm. Subsequent investigations found that the richness of native plants was (sometimes) reduced (a little) in sedge stands, but that the only species that seemed particularly affected was marram. To make the case against the sedge, researchers were careful to compare only 'sedge vs nothing', since marram itself would certainly have been found to have a similar effect on diversity. Further work showed that there's no hope of eradicating the sedge anyway; repeated applications of glyphosate were both ineffective and (because of the care needed to avoid native species) expensive.

At least the experiments that showed that spraying sand sedge with glyphosate was a waste of time did not themselves appear to damage any native species. Once spraying begins in earnest, this can no longer be guaranteed. In one study in New South Wales, drift from spraying bitou bush (*Chrysanthemoides monilifera* ssp. *rotundata*) accidentally caused serious damage to a small population of an endangered shrub, *Pimelea spicata*. Bitou bush is supposed to be one of the main threats to the survival of *Pimelea*, but subsequent investigation revealed that *Pimelea* (and several other native species) were more sensitive to glyphosate than bitou bush, so this seems very likely to be a recurrent problem.

BIOLOGICAL CONTROL AND A TALE OF TWO SNAILS

The examples we've looked at so far have involved trapping, shooting, poisoning, spraying or weeding the unwanted introduction, but often attempts at control involve the deliberate introduction of a disease, parasite or predator to control a previous introduction. Many early introductions were undertaken by people with no real understanding of the likely consequences, and often involved generalist predators that wouldn't even be considered for biological control today. It's unlikely that the small group of farmers who introduced the Indian mongoose to Hawaii in 1883 to control rats in sugarcane had any idea what a disaster this would prove to be for native birds. Introductions to Hawaii were completely unregulated before 1890, but even some later introductions didn't work out as planned. The wasp *Eupelmus cushmani* was introduced in 1934 to control pepper weevil, *Anthonomus eugenii*, an important pest of peppers. It's not easy to work out what was going through the minds of those behind the introduction, but *Eupelmus* turns out to have a huge host range. In particular it attacks *Procecidochares alani*, a stem-galling fly introduced from Mexico to help to control a range of alien weeds.

Even successful biological control agents can come back to bite you in unexpected ways. The Argentinean moth *Cactoblastis cactorum* was famously introduced around the world to control prickly pear cactus, most notably in Australia. But after being deliberately introduced to the Caribbean, it then found its way (accidentally) to the USA, where it threatens a number of native *Opuntia* species, including the endangered Florida semaphore cactus, *O. corallicola*. As it spreads, *Cactoblastis* looks likely to threaten agriculture in Mexico, the massive US ornamental cactus industry, and even an endangered iguana in the Bahamas, perhaps in the end causing more trouble than the cactus it attacks. Attention is now turning to the introduction of South American predators and parasites in an attempt to control the moth, a course that will hopefully not have the drastic impact of the attempt to control the giant African snail in the Pacific islands.

The giant African snail, *Achatina fulica*, is one of nature's great success stories. Its spread around the world from its home in East Africa has been both deliberate and accidental, but mainly the former. It's eaten by humans (young snails are sold as a substitute for edible snails, *Helix* spp.) and sometimes considered to have medicinal value. Duck farmers have often introduced *Achatina* as food for their fowl and it followed the Japanese army around the Pacific during the Second World War.

So why is *Achatina* a problem (bearing in mind it is routinely listed as one of the world's top 100 invasive species)? Well, it's very big and has an appetite to match, so it's widely assumed to be a threat to agriculture – and the visual impact of an invasion is remarkable, with thousands of alarmingly big snails apparently covering the vegetation as far as the eye can see. Unsurprisingly, this soon generates panic and, once the media get on board, the snails become a national emergency. However, reports of actual damage are scarce, and nor does the snail seem to represent a threat to human health. *Achatina* is an intermediate host of a nematode responsible for a form of meningitis, which parasitises the respiratory system of rats, which in turn can lead to transmission to humans. However, there's no link between the prevalence of the disease and the spread of *Achatina*.

A bewildering variety of agents have been considered as biological controls on *Achatina*, including beetles, flies, crustaceans and birds. The chief agent, however, has been a predatory snail from Florida called *Euglandina rosea*. *Euglandina* has been introduced almost everywhere *Achatina* is found, despite sparse evidence that it has ever had much effect on the numbers of its supposed prey. What is not in doubt is that, left to itself, after a period of rapid expansion (which is what causes the panic), *Achatina* populations always stabilise and then decline, and that this happens whether *Euglandina* is present or not. On invaded Pacific islands, it's widely believed that *Euglandina* is responsible for the decline. But because both snails are big and predation tends to be messily visible, it's easy to associate any decline in the giant African snail with predation by *Euglandina*. However, there is no evidence that this is true, and in fact plenty of other evidence that disease is a much more likely cause.

All in all, *Achatina* could be described as fairly harmless, something that cannot be said of *Euglandina*. *Euglandina* isn't very keen on eating *Achatina*, but it's an extremely effective predator of other, smaller snails. The Pacific islands support (or used to support) a staggering diversity of snails: 931 species in the Hawaiian archipelago alone. Many species had been lost

Euglandina rosea – bad news for the Pacific's snail species.

already through deforestation, but *Euglandina* soon munched its way through the survivors. On Oahu, 36 subspecies and 60 races of *Achatinella mustelina* were being actively studied when *Euglandina* arrived. All disappeared. In the Society Islands, hundreds of species in 15 families are in danger; 12 of these families have never been the subject of any detailed research, and now they probably never will. On Moorea, one of the Society Islands, a large international team studying the evolutionary radiation of *Partula* snails happened to be on hand to witness the disappearance of the entire genus on the island. In the Japanese Bonin Islands, 60 endemic species are in danger, and in New Caledonia it's predicted that the entire native snail fauna will probably disappear within a century.

The *Euglandina/Achatina* affair is exhibit A in any account of how not to run a biological control programme. Routine precautions and warnings from eminent experts were ignored, even when the danger was clear: the US Department of Agriculture introduced *Euglandina* to American Samoa (twice) even after the dire consequences elsewhere were well known. Since it's far from clear that *Achatina* actually needed to be controlled anyway, and in any case *Euglandina* failed to do so, the numerous introductions were a fiasco; the partiality of *Euglandina* to every other snail that crossed its path turned that fiasco into a disaster. Oh yes, I nearly forgot one last thing. *Euglandina* also turns out to be a much more effective vector of the meningitis nematode than poor old *Achatina*.

Sometimes the unintended ramifications of a biological control introduction are so complex, just thinking about them makes your head hurt. Consider the attempt to control European spotted knapweed (*Centaurea stoebe*) in North America, using two European flies (*Urophora* spp.) that lay their eggs in the flower heads of the knapweed and massively reduce its seed production (although not enough to actually control the weed). One reason control was ineffective is that the fly larvae rapidly became a favoured food of native deer mice. This extra food meant that the population of deer mice increased and as a result more of them became infected by a hantavirus (that also infects humans). At the same time, as the mice ate knapweed seed heads containing fly larvae, they also inadvertently consumed whole knapweed seeds that they then deposited in their dung, thus providing a novel dispersal mechanism for the weed. Not only that, but knapweed seeds eventually turned up in the faecal pellets of great horned owls that had presumably eaten deer mice, thus providing long-distance dispersal far in excess of anything the weed might have achieved on its own. Finally, if you're still with me, deer mice are actually big seed-eaters, but don't like the taste of knapweed seeds. So the greater numbers of deer mice (caused by a diet of nutritious fly larvae) ate more seeds of native plants, reducing their establishment from seed.

ALIENS AND THE LAW

Of course none of this activity takes place in isolation, and it's often a consequence of some kind of regulatory framework. This varies a lot from country to country, but England and Scotland make an interesting comparison. The legislative instrument designed to protect the citizens of England and Wales from alien plants and animals is the Wildlife and Countryside Act. Schedule 9 of the Act lists the species that cannot legally be introduced into the wild. Over the years, Schedule 9 has grown, and it recently grew quite a bit. For example, it now includes several species of *Cotoneaster* popular with gardeners, despite DEFRA (the Department for Environment, Food and Rural Affairs) noting (apparently with a straight face): 'There is little value in listing species which spread solely through natural means upon which the legislation can have little impact.' The listed cotoneasters are, of course, spread far and wide by birds, none of which can be persuaded to read Schedule 9. As the plant conservation charity Plantlife note: 'the law does not prohibit growing these plants in your garden but no gardener will be able to prevent the wind and birds from carrying seeds from gardens into the wild, and it is not clear how natural seed dispersal will be interpreted by law'. Thus one could argue that for many of these plants 'causing to grow in the wild' is implicit in growing them at all.

Another problem with prohibiting introduction into the 'wild' is that there is apparently no definition in law of 'wild'. The England and Wales legislation seems to take 'the wild' to encompass all natural habitats, agricultural and forestry landscapes, and even urban parks. It excludes only 'secure enclosures containing artificial environments'. Greenhouses, perhaps?

Meanwhile, keen to demonstrate their independence from events south of the border, the Scottish Government has brought its collective wisdom to bear on the perceived ineffectiveness of current invasive species legislation. The Wildlife and Natural Environment (Scotland) Act 2011 takes a remarkably zero-tolerance approach to alien species. Plants and animals cannot be introduced outside their native range ('the locality to which the animal or plant of that type is indigenous') and 'if plants and animals have arrived here due to human action alone, they can *never* be considered "native".' Thus rabbits, introduced by the Normans, are defined as aliens in Scotland, as are extinct natives, such as wolves and beaver. So you can do everything in your power to prevent a native becoming extinct in Scotland, but, once it does, *re*introducing it without official permission is a crime.

But what if the 'native range' is changing, as in practice native ranges always are? According to the Scottish law, 'If a range is increased naturally (for example, in response to climate change) then this larger area will be considered to be the native range of the animal or plant. However, if the range is only expanding as a result of human activity then this will not be considered the animal or plant's native range.' Thus humans are not part of 'nature' at all (although, curiously, climate change is 'natural'), and the taint of human-assisted migration makes anything non-native by definition. The Scottish Act also contains a provision to ban the sale or even the keeping of specified non-native animals or plants. No species is currently listed, but the power is there, just in case.

This Calvinist approach to alien species seems to occupy a curious twilight zone that exists only inside the Scottish Parliament. Species deemed to be native to Scotland, and which exist there in 2011, *belong* there, and nothing else does. But outside, in the real world, plants and animals inhabit an almost entirely man-made landscape, where humans are either the primary dispersers, or at the very least human land use determines what can easily migrate and what will have trouble moving at all. The idea that species can expand their ranges without human assistance, or at least connivance, is fanciful. How is it possible in the twenty-first century, even in principle, to demonstrate that a species has spread to Scotland without human assistance?

Scotland, of course, is far from alone in its absolutist attitude to introduced species. In California, the state Department of Parks and Recreation required all state parks to replace exotic plant species that might be capable of naturalising with native or 'non-invading' species. This policy doesn't exactly make life simple for park managers. Policy calls for the eradication of all exotic species, yet extermination of even one widespread and abundant plant species can rapidly exceed available funds. To see the problems this can cause, consider the attempt by the Parks Department to remove *Eucalyptus* from Angel Island State Park (the original trees were planted by the government, as usual).

For a start, why *Eucalyptus*? Of the 674 alien, naturalised species present in California at the time, *Eucalyptus* was singled out not from an objective study of environmental priorities, but because private loggers were happy to take them away for 10 cents per tonne. Shortage of funds often means targeting species that are easiest and/or cheapest to remove. From the start the assault on *Eucalyptus* was controversial. There are many other non-native species on the island, and removing *Eucalyptus* might easily lead to their replacement by other aliens such as broom, less attractive and both difficult and expensive to control. Then there was the question of the effects on native wildlife. Over half of the bird species on the island were found in both *Eucalyptus* and native oak woodlands, and some were found only in the *Eucalyptus*; total bird abundance was the same in the two types of woodland. Populations of the commonest native salamander were three times higher in eucalypt woodland. Monarch butterflies overwinter at two sites on the island, both in *Eucalyptus* groves. In fact, nearly three-quarters of all such butterfly roosts in coastal California were in *Eucalyptus*, probably owing to its provision of both nectar and evergreen shelter.

Finally, a survey of visitors to the island revealed that 98 per cent opposed the removal of the eucalypts. Many simply objected to the felling of any trees, while others actively liked their shade, appearance and smell. Not for the first time, 'informed' opinion of experts, based on geographical origin, was in direct opposition to the views of the general public, who saw things simply in terms of amenity. A compromise, involving selective felling of the eucalyptus a few at a time to create gaps in which oaks could be established, was deemed too expensive: commercial loggers couldn't make a profit and the task of felling would fall to the park authorities, who couldn't afford it.

The difficulty here is the inevitable collision between the law, which deals in black and white absolutes, and ecology, where a grey fog predominates. As US ecologist Walter Westman put it:

> While biologists have come to recognize the continually varying nature of plant assemblages, public policy reflects an older, more rigid view of community organization in which each invader was seen as representing an equal threat to host community integrity.

We often don't know what's native and what's alien, and our definitions of both are often bent out of shape to accommodate our likes and dislikes. We would like to be able to keep out harmful aliens, but we're poor at measuring harm, and reluctant to accept that most successful aliens are merely exploiting man-made opportunities, and often on the receiving end of substantial human assistance too.

Biological control is just another invasion, albeit one we are trying to encourage rather than prevent, and its frequent failure is another example of how poorly we understand the effects of adding new species to ecosystems. Only about one in three species introduced as biological controls establish at all, and only half of those that do establish (i.e. about 16 per cent of total attempts) control the intended enemy. Recall the two flies introduced to control the European spotted knapweed in North America. Even though the flies did everything they were expected to do, and even though both are model citizens (that is, they didn't attack anything apart from their intended victim), the net result was (a) no control of the weed, (b) better weed dispersal, (c) an increased disease threat to humans and (d) reduced seedling establishment of native plants. Result, eh?

Yet, despite all the evidence to the contrary, politicians persist in behaving as though ecologists (i.e. people like me) actually understand how ecosystems work (which we sort of do) and can predict their future behaviour (which we can't). Beliefs that would be hilarious if they weren't so tragic.

CHAPTER NINE

NO GOING
BACK

The world has been permanently changed by the establishment of thousands of introduced species, and except in a tiny minority of cases (mainly on small islands), eradication is not a realistic option. The aliens are here to stay. Given that, the best option in many cases may be to shift the focus away from eradication and restoration, and move to a more conciliatory approach that recognises that many alien species perform useful functions. In any case, an aggressive 'agricultural' attitude to introduced species, based on chemical and/or mechanical control and the philosophy that the only good alien is a dead alien, often simply creates more of the conditions that encouraged the invasion in the first place. As usual, Rachel Carson had it right over 50 years ago in *Silent Spring*: 'By their very nature chemical controls are self-defeating, for they have been devised and applied without taking into account the complex biological systems against which they have been blindly hurled.'

MAKING THE BEST OF ALIENS

By way of an example of a more enlightened approach, consider *Psidium guajava* (guava), a tropical tree that has been widely introduced outside its native South and central America and is regarded almost everywhere as a successful and unwelcome pest. Despite this reputation, in Kenya guava has real potential as a tool in the restoration of tropical forest. Studies of isolated guava trees in farmland showed that they were extremely attractive to a wide range of fruit-eating birds. In the course of visiting them, birds dropped seeds beneath the guavas, many of them from trees in nearby fragments of rainforest, and many of these seeds germinated and grew into young trees. Surprisingly, distance to the nearest forest didn't seem to matter at all – trees up to 2 km away (the longest distance studied) were just as good as trees much nearer to forest fragments. Guavas establish easily on degraded land, and each tree is potentially the nucleus of a patch of regenerating rainforest. Of course, most seedlings that grow beneath guavas are just more guavas, but guava is an early-successional tree that soon dies out when overtopped by bigger trees, nor does it actively invade primary forest.

Invasive alien trees can also be useful for restoring native forest. In Puerto Rico, native pioneer trees could cope with natural disturbances such as drought, hurricanes, floods and landslides, but are mostly unable to colonise land that has undergone deforestation, extended agricultural use and eventual abandonment. In these sites, low-diversity pioneer communities of invasive trees develop, but over time native trees invade. Alien pioneers may dominate for 30 to 40 years but the eventual outcome, after 60 to 80 years, is a diverse mixture of native and alien species, but with a majority of native species. In the absence of the initial alien colonists, abandoned agricultural land tends to become pasture and remain that way almost indefinitely.

154

Guava – a tropical pest or a restorer of rainforest?

A common feature of both these examples, and of many similar instances, is the difficulty we humans seem to have in thinking on the right timescale. As abandoned land in Puerto Rico is colonised by a virtual monoculture of alien trees, and as the tenure of this alien blanket extends to 20, 30 or even 40 years, the harder it becomes to resist the conviction that something must be done. In the face of mounting clamour from conservationists, local residents and the media, a policy of patient optimism becomes harder and harder to sustain. Yet this is the right policy, because on the timescale of the development of mature forests, even if not that of local politics or conservation funding programmes, 30 or 40 years is no more than a moment. The end product, a mixed forest of native and alien trees (but with a majority of natives) may not be what you might have wished for in an ideal world, but it's a whole lot better than anything likely to come from any attempt to interfere.

Time and again this turns out to be the case. In New Zealand, impenetrable thickets of invasive gorse, which must at first sight look like they will last for ever, evolve after 30 to 40 years (as long as you leave them alone) into forest dominated by native trees. Admittedly, the succession is not quite the same as that which develops from the usual native pioneer, kanuka (*Kunzea ericoides*), at least in the early stages. But these are early days; no one has observed these successions for more than 50 to 60 years, and in any case, as on Puerto Rico, the result of doing nothing is almost certainly better than that of any ill-judged intervention. Those who judge the result of the gorse succession to be inadequate recommend that 'Manually establishing patches of kanuka and manuka within landscapes dominated by gorse or other naturalised shrubs may be necessary', but wisely refrain from commenting on the difficulty and expense of doing this. Given that gorse quickly establishes a dense and persistent soil seed bank, any attempt to establish *anything* in the middle of a patch of gorse is almost certain to give rise to nothing other than more gorse.

A LONGER PERSPECTIVE

Keeping your head when all about you are losing theirs isn't always the easiest strategy to adopt, and doesn't often win you any prizes either, but it's always worth a try. If you look at second-growth forests in Ohio, all developed on abandoned agricultural land, it's clear that in the early stages species from Europe and Asia, including Japanese honeysuckle, bindweed, wild carrot and multiflora rose, are both frequent and abundant. But as succession proceeds, all these species become less common. Few remain after 60 years, and after 140 to 160 years not only have almost all disappeared, the few survivors are distinctly uncommon and it's not certain that any of them have a long-term place in mature forest. In other words, the best way to get rid of these species, at least if native forest is your ultimate aim, is to ignore them.

The eventual demise of early-successional alien plants in mature forest could at least have been foreseen, but sometimes everyone is surprised by the consequences of doing nothing. The Argentine ant is widely regarded as one of the world's worst introduced species, and has spread from South America to most of the warmer areas of the world. Argentine ants form genetically uniform 'supercolonies' that can occupy enormous areas; the European supercolony is 6,000 km across. In fact this particular genetic strain has been identified as the most successful worldwide, and also occurs in California, Australia, New Zealand, Hawaii and Japan. New Zealand is towards the cooler end of the ant's potential distribution, but it still has the potential to occupy much of the North Island and the warmer parts of the South Island. Moreover, climate models show that more of New Zealand will become favourable as the climate warms. Thus, although the ant arrived only in 1990 (from Australia), there is widespread fear that it will eventually become a major problem; a 2010 study states 'that Argentine ants are still only at the beginning of their invasion in New Zealand, and that estimated treatment costs are set to greatly increase over the next twenty years', with projected eventual annual control costs of $68 million.

Which makes a 2012 study of Argentine ants in New Zealand all the more surprising. Researchers looked for 150 established infestations across both islands, and found that 40 per cent of them had completely disappeared. Not only that, but many of the surviving populations had dwindled from numerous nests covering many hectares to just one or two nests covering only a few square metres. No one has a clue why this has happened; certainly no human effort at control is responsible. Maybe some parasite or pathogen has caught up with them, as sometimes happens in species with low genetic diversity. But the Argentine ant's peculiar biology means it always has low genetic diversity, even in its native range. On this occasion, as on many others, we just have to admit that we're poor at understanding why some aliens succeed and others fail, and even worse at figuring out why some succeed and *then* fail.

Even on Hawaii, widely regarded as the most invaded place on the planet, not everything is working out as we expected. Twenty years ago, exotic grass *Melinis minutiflora* seemed to be sweeping everything before it, setting in motion a destructive cycle of increased nutrient cycling and fire frequency which in turn led to yet more grass, a cycle that was confidently predicted to turn everywhere into a monoculture of *Melinis*. But such 'positive feedback loops' often contain within them the seeds of their own destruction, and this one is no exception. Researchers who returned to Hawaii in 2010-2012 to check on the progress of the invasion found that *Melinis* is now in decline, leading to re-establishment of woody plants. Currently, the main beneficiary of the decline is an exotic tree, but the problem for native trees seems only to be poor dispersal, suggesting that some fairly modest intervention would allow native trees to recolonise the former – and apparently permanent – grassy monoculture.

In South Africa, Argentine ants feed on the fatty elaiosome at the tip of the seed of a fynbos shrub and discard the seeds without burying them. Bad news for the fynbos community, but the ants may not be around forever.

On a longer timescale, palaeoecologists are sometimes frustrated by the short-term perspectives of all ecologists, and not just those who happen to work on invasive species. For example, most of us are aware of the precipitous declines in trees such as elm and chestnut caused by various pests and diseases, but not everyone knows that exactly the same happened to eastern hemlock around 5,000 years ago; for 1,000 years after that, hemlock virtually disappeared from eastern North America. Nobody knows exactly what happened, but the hemlock looper caterpillar is the prime suspect, although climate change may have been involved too. Palaeoecologists naturally think in terms of millennial timescales and tend not to be very impressed by how the world happens to look right now. Instead they see species responding in their own individual ways to environmental change, constantly coming together and separating into what ecologists like to call communities, but which in reality are only stills from a movie. Pause the movie a few millennia later, or even a few decades or years, and you get a different snapshot. At some scale, all species are invaders and all ecosystems are novel.

ALIEN EVOLUTION

We have seen plenty of reasons why simply trying to rewind the clock to a pre-invasion past may be difficult or impossible, and sometimes even undesirable. But there's another, more subtle reason why we can't do that. Ecologists and conservationists, and especially those with the job of drafting and implementing plans to control invasive species, generally treat those species as though they were fixed entities. But they aren't; they are constantly evolving. In fact, theory suggests that invasive aliens, with rapidly expanding populations and often starting with an unusual, depauperate complement of genes, are likely to undergo particularly rapid evolution. Increasingly we are able to observe this evolution by looking directly at genes themselves, but usually the evolutionary changes are obvious.

When sheep were first taken to New Zealand, red clover was taken for them to eat. At first the clover didn't do very well, because New Zealand lacks suitable pollinators, so bumble-bees were imported from England. To overcome the technical difficulty of transport, hibernating queens were dug up and transported while still in hibernation. The two successful intro-duction attempts involved 93 bumblebee queens in 1885 and a further 143 queens in 1906.

The queens were a random selection of British bee species at the time, and four species were successful. One of them was *Bombus subterraneus*, which was relatively uncommon in Britain even then, and has since become locally extinct. It's impossible to tell exactly how many queens gave rise to the currently thriving New Zealand population of *B. subterraneus*, but genetic analysis suggests a best guess of just two. From such unpromising beginnings it's astonishing the species established in New Zealand at all. But given this tiny inoculum, followed by over a century of separate evolution, it's less surprising that the present New Zealand population is both genetically very different from British bees and much less genetically diverse (as judged from museum specimens).

The practical importance of this story is that the New Zealand bees have been the favoured source for an attempted reintroduction of *B. subterraneus* into Britain. However the genetic analysis shows that if the aim is bees as close as possible to the original British population, the New Zealand bees would be a poor choice, even though they are the direct descendants of British bees. Under the circumstances it's perhaps just as well that raising New Zealand bees in captivity (a necessary first step in the reintroduction) has turned out to be more or less impossible, so sourcing bees from New Zealand has been abandoned. The new plan uses bees from Sweden, which not only makes the reintroduction much simpler, but they also turn out to be genetically much closer to the original British bees.

Some colonial plant introductions have also evolved remarkably quickly. When alien plants were introduced to Australia, they experienced novel climates, competitors and herbivores. If we look at species introduced a century or more ago, morphological changes can be followed by looking at herbarium specimens collected at various times since they were introduced. Among a sample of 23 introduced species, many had changed in height, leaf area or leaf shape. The biggest and most frequent changes were in height, with most plants getting smaller over time, but one or two became taller. Some species halved in height over a century, most likely an adaptive response to a drier climate. Sometimes evolution, coupled with hybridisation, leads rapidly to what is effectively a new species. Two European species of radish (*Raphanus*) have been introduced to California: 'wild' radish (*R. raphanistrum*) and the crop radish *R. sativus*. The former became a weed, just as it is in its native habitat. But the two radishes hybridised to produce a new 'feral' radish that now outperforms either parent and has completely replaced wild radish as the major weed.

EVOLUTION OF THE INVADED

Nor does the invaded community stand still; it, too, evolves in response to the invaders. Generally this evolution makes the natives better able to compete with the invaders, better able to exploit introduced prey or avoid introduced predators, or sometimes just reduces the probability of harmful encounters between resident and introduced species. For example, cane toads (*Bufo marinus*) are a notoriously troublesome invader in Australia, partly because they contain toxins novel to Australia, which has no native toads. Since they invaded, the toads have measurably evolved, becoming both bigger and more toxic, but native species have also evolved in response to the toads. Native Australian black snakes (*Pseudechis porphyriacus*) have a particular problem with the toads because they try to eat them and are killed by the toxin. In response the snakes have both become less likely to attack toads and evolved increased resistance to the toxin. But most remarkably they have also evolved smaller heads, which makes them less likely to try to eat large prey items like toads. Cane toads are still spreading in Australia and the snakes show a predictable pattern of evolution, with resistance to toad toxin increasing with the time the snakes have been in contact with them.

Maggots of the native North American tephritid fruit fly, *Rhagoletis pomonella*, eat hawthorn fruits, but some time in the mid-nineteenth century a population of the fly evolved to exploit introduced, domesticated apples. Because apples ripen earlier than haws, the new fly type has a different seasonal phenology than its ancestor. Because of these differences in timing and because *Rhagoletis* flies meet other flies and mate on their host fruits, the new and old fly types are effectively isolated from each other. At the moment, the apple and hawthorn flies are just races of the original fly, but they are destined in time to become separate species. Meanwhile two other *Rhagoletis* species have been up to something even more surprising. Honeysuckles (*Lonicera*) are hosts for *Rhagoletis* in Asia and Europe, but until 1997 no one had ever observed *Rhagoletis* attacking *Lonicera* in America. But two native American species, the blueberry fly (*R. mendax*) and the snowberry fly (*R. zephyria*) have hybridised to produce a new species (the '*Lonicera* fly') that attacks a range of introduced honeysuckles from Asia, including Morrow's honeysuckle (*L. morrowii*), *L.* × *bella* (a hybrid of *morrowii* and *tatarica*) and *L.* × *amoena* (a hybrid of *korolkowii* and *tatarica*).

Russian knapweed (*Acroptilon repens*) has invaded North America from the Caucasus and is widely regarded as a troublesome weed. In the early stages of invasion the knapweed often forms virtual monocultures, but later it seems to coexist with some native bunchgrasses such as *Sporobolus airoides*. Experiments show that *Sporobolus* with a long history of exposure to the knapweed have an increased ability to grow and survive in knapweed stands, compared to *Sporobolus* that has never met the weed.

The alpine lakes of the Sierra Nevada in eastern California, like all freshwater lakes, are inhabited by species of *Daphnia* (water fleas). The characteristic Sierra Nevada *Daphnia* is a particularly large species, which makes it attractive to fish predators. Originally this wasn't a problem because the lakes didn't contain fish, but most have been stocked with non-native trout at various times during the past century. In some lakes this led to the rapid extinction of the large, tasty *Daphnia*, but sometimes fish and water fleas managed to coexist. Where this happened, the *Daphnia* have evolved a smaller size and earlier maturity, both well-known anti-predator changes that have also been observed in other aquatic systems where prey (e.g. guppies) suddenly find themselves exposed to fish predators.

Few examples of an evolutionary response to an invader are as well documented as that of the soapberry bug, described in a series of papers by US ecologist Scott Carroll. The soapberry bug (*Jadera haematoloma*) feeds exclusively on the seeds of plants in the family Sapindaceae. Like all true bugs (Hemiptera), it has tubular piercing mouthparts ('beak'), with which it pierces the fruit wall and the central seeds, which are then liquefied and sucked up. In its native south-eastern USA, it feeds on the soapberry tree (*Sapindus saponaria*), the serjania vine (*Serjania brachycarpa*) in southern Texas, and the balloon vine (*Cardiospermum corindum*) in southern Florida. Soapberry bug races that feed on species with different-sized fruits already have beaks of different lengths, consistent with the fruit sizes of their hosts. Three other Sapindaceae have been introduced to the United States and colonised by the soapberry bug: the 'round-podded' golden rain tree (*Koelreuteria paniculata*) from east Asia and the 'flat-podded' golden rain tree (*Koelreuteria elegans*) from south-east Asia are grown as ornamentals, and the heartseed vine (*Cardiospermum halicacabum*) is a weed in Louisiana and Mississippi. Beak lengths of bugs have changed in response to

The soapberry (or red-shouldered) bug, *Jadera haematoloma.*

fruit size of the new hosts. For example, in Florida, beak lengths are much shorter in populations on the introduced tree, which has much smaller fruits than the native vine. In Oklahoma and Louisiana, beak lengths are longer in populations on the two introduced hosts, both of which have larger fruits than the native host in that region. Museum specimens allow the change in beak length to be followed, and matched to the times when the new hosts were introduced and spread.

The American soapberry bug story is interesting enough, but there's *another* soapberry bug story on the other side of the world. Australia has its own native soapberry bug (*Leptocoris tagalicus*), which feeds on the relatively small fruits of trees in the genus *Alectryon* (wild rambutans), but has colonised the Neotropical balloon vine (*Cardiospermum grandiflorum*). As its name suggests, the balloon vine has much larger fruits, and the bug has evolved a much longer beak in response. Balloon vine has really only become common in Australia since 1965, and again museum specimens chart the change in beak length since then. The new improved bug is already twice as effective as its ancestor at eating balloon vine seeds, but all the evidence suggests it could (and in time will) evolve a still longer beak and do even better.

THE TIP OF THE ICEBERG

The study of evolutionary change in invaded ecosystems is relatively new, but it's probably fair to say that evolutionary changes in invaders and in the natives they meet have been found whenever anyone has taken the trouble to look for them. But what we've found so far is the tip of the iceberg: the response of a handful of native plants to a new competitor, or a native herbivore to a new predator or potential food plant. In a word, we've noticed the things you would have expected us to notice: relatively obvious changes in a few of the more conspicuous components of ecosystems. But in reality every part of an ecosystem is changed by invaders, including all those bacteria, fungi, nematodes and other soil animals that no one takes much notice of. We don't notice these changes, partly because we don't look, and partly because we didn't have much idea of what they were like in the first place.

Thus invaded ecosystems are different, not only in the superficial sense that they consist of novel combinations of species with new and often quite unexpected interactions between them, but also in the much deeper sense that even the players that look the same are not. It's difficult, even in theory, to imagine that these changes could be rewound back to some previous 'pristine' state, even if we ignore the fact that such a state is itself largely a figment of our collective imaginations.

There's also one final important lesson for our attempts to control invasive species. As we've seen, such attempts are often pretty unsuccessful, but the more successful they are, the more they delay the inevitable evolutionary accommodation between invader and invaded. There is little doubt that such an accommodation eventually occurs. One of several searches for evidence of 'enemy release' (see Chapter 7) in invasive alien plants found, as usual, that this evidence is rather thin. But, crucially, this study also looked at time since introduction of the aliens and found a powerful effect: the best evidence for enemy release is in recently introduced species and declines rapidly with time since introduction. Depending on exactly what you choose to measure, no measurable effect remains after somewhere between 50 and 200 years, which is presumably the time it takes native herbivores and pathogens to catch up with the novel defences presented by an alien plant.

CHAPTER TEN

LEVELLING THE PLAYING FIELD

Huge numbers of plants and animals have been moved around the globe by humans, either deliberately or accidentally. Owing to increases in the speed of travel and the volume of traffic, of both people and freight, the rate of introductions shows no sign of slowing. Historically, many introductions were deliberate, and often for reasons that now seem frankly bizarre. For example, America owes its 200 million starlings to the American Acclimatization Society which, under the chairmanship of New York pharmacist Eugene Schieffelin (eccentric or nutcase, depending on your opinion), set out to introduce to America every bird species mentioned in the plays of Shakespeare. Shakespeare makes about 600 mentions of 50 different birds in his plays. Eagles are top, and starlings appear only once, in *Henry IV, Part 1*. Henry has refused to pay a ransom for Hotspur's brother-in-law Mortimer, and has forbidden Hotspur to mention his name. To keep Mortimer's name before the King without actually disobeying this instruction, Hotspur contemplates teaching a starling to repeat the name: 'Nay, I'll have a starling shall be taught to speak nothing but "Mortimer," and give it him, to keep his anger still in motion.' So just the one mention, but it was enough to get 100 starlings released into Central Park, and the rest, as they say, is history.

DELIBERATE INTRODUCTIONS: THE STRANGE TALE OF THE HARLEQUIN LADYBIRD

Harlequin ladybirds do not feature in Shakespeare, but were another deliberate introduction to the USA. They are interesting on two scores: first, as an example of how even a planned introduction may rely on a large slice of luck; second as an illustration of just how unpredictable successful introductions can be. With hindsight we can work out what happened, but no one could have predicted it in advance. The harlequin also illustrates yet again the power of propagule pressure; what happened to the harlequin may have been highly unlikely, but if you keep trying, even unlikely events happen eventually.

The Asian *Harmonia axyridis* (harlequin ladybird) was first recorded in Britain in 2004 and can now be found almost everywhere in the country, although it's still uncommon in Wales and rare in Scotland and the far north of England. But how did it arrive, and why is it so successful? Given its native range of China, Japan and the far east of Russia, you could be forgiven for assuming the harlequin arrived direct from Asia. But it didn't – it came from America. But it was far from an overnight success there, or anywhere else for that matter. In fact, the harlequin's history is full of surprises.

A century ago, the harlequin's voracious appetite suggested it might make a good biological control agent for a range of insect pests, and attempts to introduce it to the USA for that

The harlequin ladybird – almost certainly on a leaf near you, soon.

purpose began as long ago as 1916 and continued on and off throughout the twentieth century. Attempts to introduce it to Europe began in 1982, and to South America in 1986. These attempts were uniformly unsuccessful, but then suddenly a thriving population was discovered in Louisiana in 1988, from where it quickly spread to the rest of America and soon afterwards to most of the rest of the world. American harlequins turned up in Europe and South America in 2001, South Africa in 2004. In North America, it's now the most widespread ladybird, and, if it isn't already Britain's commonest ladybird, it probably soon will be.

Clearly something happened to the harlequin in the south-eastern USA in the 1980s, and whatever it was, it had much the same effect that being bitten by a radioactive spider had on Peter Parker: the harlequin went into Louisiana a wimp and came out bent on world domination. To understand what happened, we have to briefly consider why the harlequin was so unsuccessful for so long, and indeed why most introductions of animals to new habitats fail. Introduced populations always start out small, and a big problem for small populations is inbreeding; because there are few animals around, they end up mating with close relatives, even brothers and sisters. Inbreeding brings together deleterious recessive mutations (which normally lurk undetected in large populations), causing malformation, disease and death.

But there's a twist – geneticists had figured out that small populations might go through a process called purging, in which the harmful mutations that make inbreeding so dangerous are lost completely. Any population that successfully negotiated this process would emerge not only immune to the detrimental effects of inbreeding, but fitter, faster and better all round.

Because the conditions under which it's supposed to occur are so restrictive, purging remained merely a theoretical curiosity – until now. Work by a combined team of American and French biologists has shown beyond any doubt that the Louisiana harlequins were successfully purged of their bad mutations in the 1980s and now grow faster and have more offspring than native Asian harlequins. They're also completely immune to inbreeding – even offspring of matings between siblings show no ill effects at all – which means they can now easily establish from even a tiny starting population. Indeed, we are now in the bizarre situation where harlequins could even be invasive back in their native range, potentially able to brush aside the native population from which they evolved.

GARDENERS' WORLD

Most introductions, especially in the recent past, have been at least partly accidental, and they continue at a remarkable rate. Hawaii acquires about 20 new arthropods every year, mostly insects, and most of the time no one has a clue how they arrived. Sometimes it's even impossible to tell where they came from. Other times the route is fairly clear – and most often, these days, it's gardeners who are responsible.

Horticulture has long been one of the biggest global movers of plants, formerly mostly of species that might prove useful as crops, but recently chiefly ornamental species. Various studies demonstrate that being taken into cultivation greatly increases the probability of subsequent escape into the wild. Cultivation gets plants over that difficult initial stage of obtaining a foothold in a new country; accidental introductions have to take their chances, but we make every effort to make sure that cultivated species do not fail at that first hurdle.

For example, one of the remarkable features of the South African flora is the 1,036 species of native Iridaceae (*Iris, Crocosmia, Gladiolus,* etc.), more than half the global total. Many Iridaceae have attractive flowers, and nearly a third of the South African species have been introduced to other countries as garden plants. Sixty-seven species of South African Iridaceae have become naturalised elsewhere, and a few have gone on to become quite troublesome weeds, especially in Australia. All of these 67 species are in cultivation; in other words, although being grown in gardens doesn't guarantee that a species will escape (many cultivated species haven't, at least not yet), *not* being in cultivation *does* guarantee that it won't.

Cultivation also reveals, once again, the crucial role of propagule pressure. An analysis of old British seed catalogues shows clearly that plants that were more widely available (sold by more companies, over longer periods), and whose seeds were cheaper, were more likely to go on to escape from culti-vation and become successfully naturalised. It seems fair to assume that cheap, widely available plants were grown more often, and thus simply had more shots at escaping than did rarer, more expensive species.

Our passion for gardening also has a knock-on effect on animal introductions. New Zealand flatworms first turned up in the UK in 1963, in a couple of gardens in Belfast, but very quickly spread to Scotland. They are still much more common in the north of the UK, which fits with their distribution in New Zealand, where they are confined to the cooler and damper parts of the South Island. They have spread around Britain in much the same way they got into the country – with contaminated plants. They have a tendency to hide under objects during the day and their sticky mucus means they can easily be transported stuck to the bottoms of things like plant pots and garden ornaments. They've also been found among the roots of containerised plants, inside plant pots, in manure heaps, and stuck to the bottom of silage bales. Genetic analysis of British flatworms shows clearly that they are not the result of a single oversight; they have been introduced from New Zealand on several occasions. In warmer parts of Britain, an Australian flatworm is also well established.

Another Antipodean animal that's probably also moved around with plants by gardeners is *Arcitalitrus dorrieni*, a land-hopper or 'lawn shrimp'. Britain's only really terrestrial crus-taceans are isopods, known as woodlice, pill bugs or sow bugs. But in a few other places, amphipod crustaceans have adopted the same lifestyle, although none are enormously well adapted to life on land. Amphipods are laterally flattened and, as their name suggests, get around by hopping. European hoppers never get far from the sea, and will be familiar to anyone who has ever turned over a heap of seaweed at the beach, but in

Australia they occupy the woodlouse niche, i.e. damp leaf litter. *Arcitalitrus dorrieni*, a native of New South Wales, was first found on Tresco in the Scilly Isles, but has since spread to mainland Britain. British gardeners are still often rather taken aback by their flattened bodies and jumping abilities, which give them a superficially flea-like appearance.

Gardeners again are almost certainly responsible for the fact that three species of stick insect from New Zealand are now established in gardens in Devon, Cornwall and the Isles of Scilly (Britain has no native stick insects). They undoubtedly arrived as stowaways on tree ferns and other ornamentals. The Indian stick insect is widely sold by pet shops, but cannot survive outdoors on the British mainland. Other British invaders, such as the large European spider *Segestria florentina* and the European yellow-tailed scorpion (*Euscorpius flavicaudis*), are found close to ports, suggesting that they were also stowaways in cargo.

Modern sealed containers may have reduced the scope for this kind of dispersal, but some particular types of cargo were once very important sources of alien species. The propensity of plant seeds to stick to the fleeces of sheep meant that, wherever fleeces were traded or processed, a characteristic plant community of 'wool aliens' tended to develop. Establishment was aided by the fact that wool waste or 'shoddy' was widely used as a fertiliser in gardens and allotments. For example, a characteristic community of wool aliens still occurs near a former wool-processing factory in Brno in the Czech Republic. The plants, reflecting the origins of the wool, are mostly Australian or South American. Similar (but different) plant communities are also associated with the importation and processing of other agricultural produce, e.g. 'bird-seed aliens'.

And, just in case we need reminding that there really are no limits to the ways alien species arrive, or to human folly for that matter, how about *Phagocata woodworthi*, a freshwater flatworm from North America that is well established in Loch Ness. It arrived – and I'm sure you're ahead of me here – as a contaminant of equipment brought from America by scientists searching for the Loch Ness monster.

JAPANESE KNOTWEED: LICE TO THE RESCUE

Another horticultural introduction, Japanese knotweed (*Fallopia japonica*) is considered to be one of Britain's worst invasive aliens, and most intractible problems, as it spreads along railway lines and colonises roadsides. A national eradication programme, based on physical and chemical methods of control, would cost an estimated £1.5 billion, and even then success could not be guaranteed.

The only realistic, affordable answer is biological control, to which (providentially) Japanese knotweed is almost uniquely vulnerable. All the Japanese knotweed in Britain derives from a single introduction, and is thus a single (female) clone. Therefore it cannot reproduce by seed and has virtually no genetic variability; if challenged by a pest or disease, its chances of evolving out of trouble are approximately zero.

You do find plenty of viable seed on Japanese knotweed, but the pollen has almost always come from its close relative, Russian vine (*F. baldschuanica*). Given the bad behaviour of both parents, a Japanese knotweed–Russian vine hybrid sounds like a nightmare, but it doesn't seem to be; seedlings of the hybrid hardly ever establish in the wild and there are only seven identified hybrid individuals in the whole of Britain and Ireland. Far more problematic is the hybrid (*F.* × *bohemica*) with its very close relative, giant knotweed (*F. sachalinensis*). Despite being even bigger, giant knotweed is less of a problem than Japanese knotweed, but the hybrid is just as bad as its mother, although not nearly as common – yet.

For the last decade CABI (the Centre for Agricultural Bioscience International, based at Wallingford in Oxfordshire) have been looking for a suitable biological knotweed-killer. There is no shortage of candidate control agents. One possibility is a leaf-spot fungus, but its complex life cycle means it's difficult to work on, so attention has focused on plant-eating insects. As you would expect for such a big, juicy, common species, plenty of insects like to eat Japanese knotweed. A thorough study of published work on the plant, combined with a search of 140 sites on all four main islands of Japan, revealed 186 insects. Many looked promising at the start but quickly had to be discarded. One leaf beetle was capable of reducing Japanese knotweed plants to a mere skeleton in no time at all, but also proved to have a taste for buckwheat, a crop plant in the same family. Attention quickly narrowed to a single suspect, the psyllid *Aphalara itadori* (psyllids, or jumping plant lice, are tiny sap-sucking insects related to aphids, but look more like miniature cicadas).

The omens for *Aphalara* are good. Observations in the field showed that it could do a lot of damage to knotweed plants. It also seemed to be highly specific – despite careful searches in Japan of related plants, even when growing right next to the villain itself, *Aphalara* was only ever found on Japanese knotweed (*Fallopia japonica*), giant knotweed (*F. sachalinensis*) and the hybrid (*F.* × *bohemica*). The specific name *itadori* even means 'Japanese knotweed' in Japanese.

Despite these encouraging signs, lengthy testing was still necessary to make sure *Aphalara* wouldn't eat anything else. CABI drew up a long list of test plants, including almost everything in the same family in Britain, related ornamentals (e.g. *Rheum palmatum*), related crop plants (e.g. buckwheat and rhubarb), a selection of plants (native and introduced) in related families, unrelated plants with similar biochemistry, a few plants

Japanese knotweed – will it survive the attack of the psyllids?

that happen to look a bit like knotweed (e.g. bindweed) and, finally, a random selection of completely unrelated crop plants for good measure. The chosen plants were then exhaustively tested. Would *Aphalara* lay eggs on them? If it would, did they hatch, and did the nymphs survive? If the egg stage was bypassed and nymphs transferred directly onto the test plants, did they survive? If adults were placed onto test plants, did they survive? *Aphalara* passed all tests with flying colours; essentially it refused either to eat or lay eggs on anything except Japanese knotweed or the hybrid *F. × bohemica*. It was less keen (but did survive) on giant knotweed, and adults placed on anything else showed their displeasure by promptly expiring. There is just a faint suggestion that it might survive, although poorly, on wireplant (*Muehlenbeckia complexa*), but even this is mainly a bonus, since wireplant is proving a bit of a problem in south-west England.

Since Japanese knotweed is herbaceous, the psyllids have to find somewhere else to spend the winter, and in Japan they hide among the foliage or crevices in the bark of various conifers, including pines and Japanese cedar (*Cryptomeria japonica*). In Britain we could expect them to use a similar range of native and introduced conifers, but there's no evidence that they feed on or damage their overwintering shelter plants.

No one believes that the extraordinarily wide range of plants exposed to the psyllid during laboratory testing was really necessary, but this belt-and-braces, everything-including-the-kitchen-sink approach is another consequence of this being a first for Europe; there are no rules, so the decision was made to err on the side of caution. As the researchers themselves put it: 'some scientifically irrelevant test plant species were included to satisfy anticipated requests by future reviewers of any application and the general public'. The 'scientifi-cally irrelevant test plant species' were indeed irrelevant. Will *Aphalara* eat aubergines or broad beans? No, it will not. In other words, for 'anticipated requests' read 'stupid questions'. What the researchers are trying to tell us, as politely as they can, is that, unless galloping paranoia about introduced species can somehow be restrained in future, this may not only be the first time biological control of a weed has been attempted in Europe, it may also be the last.

Following government approval, the psyllid was released at specially chosen sites in the spring of 2010. All we know so far is that it came through the unusually cold winter of 2010–2011, but it will be years before we know the effect on the wider knotweed problem.

FELLOW TRAVELLERS

From starlings to flatworms, and wool aliens to wheat, it's hard at first sight to see what all these introduced species have in common, but in fact all do share one elusive quality: they thrive around people. For plants and animals introduced deliberately for food, as pets or for ornament, that's obvious, but it's equally true for accidental introductions. Before they could be moved by human agency, those New Zealand flatworms had to find their way into plant pots, wool aliens onto domesticated sheep, rats and brown tree snakes onto ships, and zebra mussels into the ballast water of ships. The sheer volume of modern trade and travel means that any species that likes the company of *Homo sapiens* or his buildings, vehicles, pets, gardens, livestock or fields will eventually find itself dispersed to pastures new. I'm tempted to call such species *anthropophiles*; I'm not quite sure that the word exists, but if it didn't, it does now.

At the same time that we are dispersing species in unprecedented numbers, we're also making such dispersal more and more necessary. Species have always found it useful to be able to disperse to new habitats, but habitat destruction and fragmentation make effective dispersal more vital than ever, if species are to survive. In addition, there is the growing threat of climate change, which means that everything has to move, sooner or later. How fast does this migration have to be? The instantaneous velocity of climate change varies a good deal from place to place, but has a global mean of 0.42 km yr^{-1}. That's a little over a metre a day, which doesn't sound like much, but it means that only 8 per cent of the Earth's protected areas have 'climatic lifetimes' of more than a century. In other words, in 92 per cent of reserves the temperature of the reserve's coldest part (nearest the pole) will exceed the present temperature of its warmest part (nearest the equator) in only 100 years. Of course, individual species may have climatic tolerances that allow them to survive in changed climates, or they may be able to make small migrations to different aspects or altitudes. But, for many species, confined to mountaintops or particular types of geology, surrounded by unsuitable habitats (especially urban or intensive agriculture) or facing insurmountable barriers such as water, dispersal may be difficult or impossible.

The growing crowd of anthropophiles, of course, face no such difficulties. Anthropogenic dispersal means they are probably already way ahead of where they need to be in even 100 years' time. For example, most European plants that have been taken into cultivation are grown far to the north of their natural distributions. The pretty southern European *Saponaria ocymoides* (rock soapwort) is not unusual: it can be bought in Vadsø in northern Norway (one of the most northerly settlements in the world, just round the corner from Murmansk), 2400 km north of its natural limit. Whether it will *grow* in Vadsø is another question, but you get the idea. Chusan palm (*Trachycarpus fortunei*) originates in China but has been planted right around the world and is now well established and thoroughly naturalised in southern Europe. It's also grown right up into the north and east of Europe, where it is routinely badly damaged or even killed by severe winters, but optimistic gardeners cheerfully replant, making sure that it and many species like it are poised to take advantage of any future climate warming.

Since this is a crucial but little-understood point, I want to pause briefly here and emphasise where that leaves us. We live in a world where a large number of species (but actually quite a small fraction of the world's biota) have been, and often still are, routinely moved distances both large and small. Because these species are not only moved by man, but are also (by definition) often the best equipped to take advantage of habitats modified by man, their future survival (and indeed success) are hardly in doubt. Among their ranks are those few species that keep some people awake at night, although, as we've seen, blaming the invaders themselves for this state of affairs is counterproductive and futile. On the other hand, the great majority of the world's species don't get on particularly well with mankind and all his works. As a result they tend not to be dispersed by human agency and, even if they were, they're usually poorly equipped to escape from ports, fields and gardens anyway.

ASSISTED MIGRATION

If you stop and think about the previous paragraph, a heretical idea might occur to you: why not try to move some of the most threatened 'losers' to new habitats, if that's what it will take to ensure their survival? Such translocation is not a fashionable idea, partly on account of paranoia arising from some disastrous previous introductions, both deliberate and accidental. As a result, moving species, even for the best of motives, has become increasingly expensive and laborious. Certainly the regulation of deliberate introductions seems out of proportion to the ease with which insects can be introduced accidentally. Consider on the one hand the decade of work and millions of pounds it took to get the UK's first weed-control insect introduced, and on the other hand how easy it is for insects to hitch-hike on cultivated plants.

At the conclusion of the 2011 Chelsea Flower Show, various *Aloe* plants and other succulents from the Kirstenbosch/South Africa stand were donated to the Royal Horticultural Society and moved to the Society's quarantine glasshouse at Wisley. These plants soon proved to be a veritable Noah's Ark of South African insects, yielding a mirid bug, a cushion scale, a woolly aphid, a diaspid scale and a mealybug. Some of these had been recorded before at other botanic gardens but are not widely established in the UK; the mirid bug and the cushion scale were new to Britain. These particular plants and their animal passengers ended up on the bonfire, but, with imports of ornamental plants to the UK worth over £1 billion annually, it's easy to see why the RHS records two or three new established insects in the UK every year. Note, by the way, that no one has to be careless or incompetent for this to happen; the Kirstenbosch plants were inspected by DEFRA Plant Health and Seeds Inspectors and again by RHS entomologists when the plants arrived at Wisley at the end of May. Neither inspection revealed any cause for concern, but by early July the infestations were apparent.

There's also a profound reluctance among ecologists/ conservationists to countenance species translocations on principle, even if it's not always clear exactly what the principle is. There is dark but vague talk of 'environmental or social harm' and 'scant appreciation of the cultural as well as scientific value of native biodiversity'. A paper by US invasion ecologists Anthony Ricciardi and Daniel Simberloff has the simple title 'Assisted colonization is not a viable conservation strategy'. Underlying these mutterings is the idea that things are exactly how they should be, and that any change is bad, a belief that I will return to in the next chapter. The most that the purists will allow is reintroduction of extinct species to sites within their former range – although recall the Scottish legislation that disapproves of even this.

Yet, on the ground, enlightened conservationists are laying the groundwork for assisted migration, so they can move fast when the time comes. In Finland, the threatened Siberian primrose (*Primula nutans* var. *jokelae*) grows on the shores of the northernmost extremity of the Gulf of Bothnia. Climate change may threaten this species, and if so the next suitable habitat may be on the shores of the Arctic Sea, 500 km north of its current range. But how critical is climate to its survival? And could it already grow by the Arctic Sea? To answer these questions, the University of Oulu Botanic Garden (near its present range), the University of Helsinki Botanic Garden (further south) and Svanhovd Botanical Garden in Norway (far to the north) are collaborating to grow the primrose. If the primrose expires in Helsinki, then we know what to expect when the climate warms; and if it thrives in Svanhovd, we know where it needs to go.

In New Zealand, endangered species are often established on offshore islands free from rats and other predators, even if the species concerned are not known to have occurred there previously. In a few places, faced with the stark choice of 'move it or lose it', things have gone even further. The Bermuda petrel is one of the world's rarest birds, and was in fact thought to be extinct for 300 years. When a few pairs were rediscovered nesting on tiny rocky islets in immediate danger of destruction by hurricanes and flooding, chicks were moved to a reserve safe from flooding on a larger island. Attempts are under way to establish *Torreya taxifolia*, arguably the world's most endangered conifer in its native Florida, in the cooler climate of North Carolina. The main problem in its tiny native range is a fungal pathogen, but climate change would soon make the move necessary anyway. A measure of the attitude of the conservation establishment to such endeavours is that the *Torreya* relocation is being undertaken by a group of concerned individuals and has no 'official' status.

Since translocation is clearly an option when imminent extinction is the only likely alternative, why not consider this option before such a desperate stage is reached? When considering your reaction to such a proposal, it's important not to forget that such a process is already under way on a vast scale, involving deliberate introductions of pets, crops, forest trees, garden plants and all their associated accidental passengers. A colleague recalls attending an official meeting where discussion of the regulations governing animal introductions to the UK was taking place. When talk turned to raccoons, he quickly checked on his iPad and found that for only a few hundred pounds he could be the proud owner of a pair of raccoons, no questions asked. In other words, raccoons were just another stable door waiting to be bolted.

So the question is not whether species should be moved, but *which* species should be moved. A 'no translocation' policy simply stacks the odds even further in favour of raccoons, weeds, aphids, mealybugs and all the other common, widespread, easily dispersed anthropophiles. Assisted migration of endangered species is a small step in the direction of allowing them to compete on a playing field that still slopes uphill, but slightly less steeply than before. Critics argue that moving species is 'playing God', but we're playing God already; it just happens to be a rather Old Testament sort of God, one who has taken to heart the maxim that 'the Lord helps those who help themselves', and interpreted this to mean that those who are in no position to help themselves don't deserve any help at all.

Chris Thomas, an ecologist at York University, in northern England, has suggested that Britain is ideally placed to receive refugees fleeing the warming climate of mainland Europe – his list (see facing page) provides a fascinating array of candidates. In fact Britain, which has been an island for only 8,000 years, has a flora and fauna that effectively already consists of such refugees. This happened in every recent interglacial, and in each one a different set of European plants and animals caught (or missed) the last boat as Britain once more became an island. It's only chance that rabbits and *Rhododendron ponticum* are not British natives; both were here in a previous interglacial.

Six candidates for translocation to Britain

Chris Thomas has come up with this list of candidates that could benefit from translocation to Britain. He's unlikely to see his plans realised, but this is an effective ecological blueprint for adapting to climate change.

IBERIAN LYNX, *Lynx pardinus*. The most endangered cat in the world is now confined to just two tiny areas of Andalucía, with fewer than 200 animals in total. It used to be considered a subspecies of the larger and much more widespread Eurasian lynx, *Lynx lynx*, but is now known to be a separate species, descended from a lynx that lived more widely in Europe during the Pleistocene. It has been suggested that the Eurasian lynx should be reintroduced to Britain, but establishment of the

Iberian lynx would find Britain a welcome haven, and might also contribute to keeping the rabbit population in check.

Iberian lynx would represent a much more significant contribution to world conservation. The Iberian lynx is a specialist rabbit predator, so it wouldn't have much trouble finding food.

PYRENEAN DESMAN, *Galemys pyrenaicus*. This is basically an aquatic mole, but with normal front feet, large webbed hind feet and a long tail. It is restricted to streams in the high mountains of northern and central Spain, where it is seriously threatened by climate change. There seems no reason why it wouldn't thrive in streams in western Britain. There's no evidence that the Pyrenean desman ever occurred in Britain, but its close relative the Russian desman was found in Britain in a previous interglacial.

SPANISH IMPERIAL EAGLE, *Aquila heliacea adalberti*. Another extremely rare rabbit specialist, with fewer than 500 individuals in Spain and Portugal.

PROVENCE CHALKHILL BLUE, *Polyommatus hispanus*. The Provence chalkhill blue butterfly in France from the eastern Pyrenees via Provence as far as the Alps and northwards via Ardèche to the Jura, and coastal north-western Italy as far as North Tuscany. It's not currently endangered, but its long-term survival is threatened by climate change. Southern England is, or soon will be, climatically suitable – and its larval food plant, horseshoe vetch, is common on calcareous grasslands in southern England.

DE PRUNNER'S RINGLET, *Erebia triaria*. De Prunner's ringlet is a butterfly that is endemic to the Alps and mountains of northern Spain, with small outlying populations in Croatia and Bosnia. It is threatened by climate change throughout its range, and England represents a considerable portion of its projected future range. The larvae feed on widespread grass species.

IBERIAN WATER BEETLES. Many of the 120 water beetle species endemic to the Iberian Peninsula occupy headwater streams in one, or sometimes a few, mountain ranges. They are under threat from increased droughts.

Given the large pool of potential (but unlucky) colonists just the other side of the English Channel, it's not surprising to find that natural colonisation is still under way. We looked at a couple of examples in Chapter 1, but another is the sunbleak, a small fish in the carp family, which is declining in its native central European range, but escaped or was released from aquaria in Britain in the 1980s and is now doing very well indeed. Were it not for the fact that Britain has been an island for the last 8,000 years, the sunbleak would almost certainly have made its own way here by now. The UK Environment Agency, on the grounds that sunbleak is foreign and therefore must be a threat to *something*, are making half-hearted attempts to limit its spread. In Sweden, by contrast, where sunbleak is regarded as a rare native, this inoffensive fish is seen more as a conservation issue.

British botanists continue to argue about the small-flowered tongue orchid (*Serapias parviflora*), previously known only from mainland Europe, which was found growing in Cornwall in 1989. Orchids produce vast numbers of extremely tiny, light seeds, and it's entirely possible that they blew across the Channel. If they did, then the tongue orchid arrived without human assistance, which means it qualifies as a native (and for careful protection). Then again, it's equally likely that seeds fell out of someone's trouser turn-ups or arrived stuck to the sole of someone's shoe, or possibly that it was deliberately planted. If any of those is the case, then it's just another bloody weed, to be ruthlessly exterminated.

Returning to Chris Thomas's list of candidates for translocation to a future 'lifeboat Britain', it's also worth noting that some similar candidates are already in Britain, by accident or design. For example, the Caucasian wingnut tree (*Pterocarya fraxinifolia*) grew wild in the British Isles in previous interglacials and shows every sign of doing so again, with numerous records of trees established outside gardens. *Pterocarya* is one of many candidates that are endemic to the Caucasus Mountains and to humid forests of the eastern Black Sea coast and southern Caspian, where moisture-dependent species are threatened by projected major reductions in summer rain.

The transient nature of the concept of 'native' can be appreciated from some of the other species formerly present in Britain. Among a long list that thrived in Britain in the relatively recent past, but did not make it to the present day are elephant, rhinoceros, musk ox, two species of lemming, hippopotamus, lion, spotted hyena, reindeer, bison, long-tailed ground squirrel, fallow deer, horse, European pond terrapin, fir (*Abies*), spruce (*Picea*), hemlock (*Tsuga*), firethorn (*Pyracantha*), water chestnut (*Trapa natans*), *Ephedra*, water fern (*Azolla*) and Montpellier maple (*Acer monspessulanum*). Many are now (re)introduced, but obviously not all!

Don't think I underestimate difficulties of assisted migration. And it's not the answer for all (or even most) species. But fear of invasive species shouldn't prevent us considering the idea in principle, although I suspect it will. Sometimes it seems many conservationists would prefer a species to go extinct than to survive somewhere it doesn't 'belong'.

CHAPTER ELEVEN

FIVE MYTHS ABOUT INVASIONS

S o low has the stock of introduced species fallen that it is now possible to commence a scientific paper with the line 'Many ecosystems worldwide are dominated by introduced plant species, leading to loss of biodiversity and ecosystem function' in the certain knowledge that no one will be surprised by such a statement, and certainly that no one will ask for it to be supported by any actual evidence. Of course, this text is pure boilerplate, not intended to trigger the firing of many – or even any – neurons in the reader's brain, maybe not even intended to be read at all. Nor do those who wrote it necessarily believe it; at least, not in the sense of having arrived at that conclusion by weighing the evidence. It's just something that gets routinely stuck on the front of any paper about alien species. But are these and the other stock assertions about alien invaders really true? It's time to nail some myths.

#1 ALIEN INVASIONS REDUCE BIODIVERSITY AND ECOSYSTEM FUNCTION

Fortunately, for plants we have a recent (2011) meta-analysis that provides exactly the evidence we need to assess if this statement is true. A meta-analysis is an analysis that combines several (ideally *all*) studies on a particular topic, and subjects them to a new, formal statistical analysis that arrives at an estimate of the overall size of an effect, i.e. does X have any effect on Y, and, if so, in what direction? In this case, all the available studies of the effects of alien plants were combined. So what was the answer?

Well, on biodiversity the study found that alien plants did indeed lead to big reductions in the diversity of invaded plant communities. But this part of the result was inevitable before the researchers set to work: inevitable because of the sorts of plants that invasion biologists consider worth the effort of studying. As they put it:

> In the vast majority of studies, invaded sites had high alien abundance and although the measures of plant abundance were not always given, the study sites were usually described as having high or > 50 per cent cover.

In other words, those interested in the effects of alien plants choose to look at plants that are capable of achieving high abundance (and almost certainly high biomass), of dominating the plant communities they invade and, *ipso facto*, of causing big reductions in local diversity. There's nothing surprising about this, it's just the way science funding works. When you go to your local research funding agency for some money, it's a whole lot more convincing to be able to say you want to study X because it appears to be a major threat to something.

So, what sort of plants exactly are we talking about? Japanese knotweed? Check. Pampas grass? Check. Common reed? Check? Giant reed (*Arundo donax*)? Check. Douglas fir? Check. Reedmace, or cattail (*Typha*)? Check. Cordgrass (*Spartina*)? Check. Giant hogweed? Check. I could go on, but you get the picture: these are plants that take no prisoners and at best elbow the natives out of the way.

The results were rather different in regards to ecosystem function. Of course, these are big plants, with big effects, not just on their neighbours but on everything. Ecosystems have lots of functions, and a lot depends on which ones you choose to measure. But among the long list of things that have actually been measured, most were unaffected by alien plants or – in the case of productivity, microbial activity, soil carbon, total N and P, and available nitrogen – were *increased* by the presence of alien plants. Science does not attach subjective, moral value to such results, and you could argue that the levels of these things in the uninvaded communities were 'just right', so that *any* change is negative. But leaving aside such metaphysical sleight of hand, it's hard to argue that this represents 'loss of ecosystem function'.

None of this would matter if such species were a representative sample of all introduced plants. But a 2013 study that asked exactly that question, i.e. 'Are these studies representative of alien plant impacts as a whole?', came swiftly to the conclusion 'It does not appear so.' Measuring effects on biodiversity and ecosystem function takes time and money and tends not to be done all that often. Even though the researchers were careful to include all the evidence they could find, they still had data for only 135 plant species – a small proportion of the well over 5,000 naturalised alien plants in Europe alone. Those 135 plants were chosen because the researchers thought they looked like they were having big effects – and they were right.

It's actually even more selective than that. Just *nine* species account for one-third of all quantitative studies of the ecological impacts of introduced plants. What's more, it turns out that what we think we know about even this tiny minority isn't all that reliable. The larger the number of studies undertaken on a particular species, the less likely they are to find any effect at all, and, if they do, the more likely it is to be the opposite of what we originally thought (for example, if an earlier study found a plant *increased* the rate of nutrient cycling, later ones might find the same plant *reduces* the rate).

Why is that? Simply because, when researchers start out looking at species X, in their zeal to find *something* they often begin by choosing a site where it looks like that something might be big, obvious and easy to measure. So, even if a species only occasionally achieves high cover, they naturally start out with a site that's 100 per cent X (even though it's neither surprising nor interesting to find, say, lower diversity at such a site). Later researchers, whether by accident or design, often look at sites with less X and find, naturally enough, smaller effects, or no effect, or the opposite effect. The more we look, the less we find. Or, to put it another way, we know nothing at all about the impacts of at least 99 per cent of alien plants, and next to nothing about the other 1 per cent.

Look at it this way: if we chose 135 *native* plants, and the places they grow, using the same peculiar criteria, we would get similar results. Or suppose we looked at the same 135 aliens in their native habitats; almost all of them would be found to be having exactly the same effects they had as aliens – that's just the kind of plants they are. For example, some New Zealand conservationists lie awake at night worrying about heather (*Calluna vulgaris*), but in my experience it's no more effective a monopolist in New Zealand than it is at home in the UK. Heather covers hundreds of square miles of British uplands to the virtual exclusion of everything else; in fact heather moorland is so spectacularly dull that when surveys revealed it

Bracken takes over woodland – but it's native, so no cause for alarm.

was becoming slightly more diverse, this was seen as a minor crisis. If a Martian were parachuted into Britain and asked to guess which plant looked most like it had arrived from the planet Zog and was bent on world domination, my guess is that the obvious choice would be bracken. In fact, if bracken were alien, it would be seen as no less than a national emergency, but it's native, so its bad behaviour generally goes unremarked. But the overwhelming majority of plants – native or alien – are not like bracken, or heather, or Japanese knotweed.

What would happen if we were to look at the effects of a random sample of alien plants on biodiversity and ecosystem function, and compare those effects with those of a similar random sample of natives? Such a study has never been done, but we can get an idea of what we might find by looking at that well-known invasions laboratory, Hawaii. We've already noted that large numbers of plant introductions and very few extinctions have greatly increased the richness of Hawaii's flora, and this is true at every scale: patches of invaded forest of any size have more tree species than similar patches with only natives. What does this mean for ecosystem function? Well, productivity, nutrient turnover and below ground carbon storage either don't differ, or are significantly greater, in invaded forests. But greater diversity does not lead to better functioning, in Hawaiian forests or anywhere else.

What is actually happening is that no native Hawaiian plant can fix atmospheric nitrogen, but some of the invaders can, so there are more nutrients in invaded forests. That makes Hawaii a bit of a special case; normally if we were to add a few exotic species, or randomly replace a few natives with aliens, most likely we would find … nothing. That is, plants are just, well, plants, and we find unusual effects only if we single out unusual plants (and often, only when growing at unusually high densities). So the sentence at the start of this chapter ('Many ecosystems worldwide are dominated by introduced plant species, leading to loss of biodiversity and ecosystem function') is both partially correct and at the same time wholly misleading. Dominance by a tiny minority of highly unrepresentative introduced plant species does lead to loss of (small-scale) biodiversity, but generally also to increased ecosystem function. But exactly the same is true of a tiny minority of natives. Most plants – alien or native – do neither of these things.

#2 ALIEN SPECIES COST US A FORTUNE

We are frequently informed that the economic costs of alien species are large, but it's quite hard to find out how large, or where the quoted figures come from. Where better, though, to find some answers than the second edition of one of the standard works on the subject, *Biological Invasions: Economic and Environmental Costs of Alien Plant, Animal, and Microbe Species.* According to its US publishers, this book 'represents the most current, single-source reference containing scientific and economic information on this timely subject'.

Most of the book's chapters deal only with one group of invaders (alien weeds in New Zealand, alien vertebrates in Europe, etc.), but one chapter tries to cover all invasive species in the USA, so let's take a look at that. At the outset, this notes, without much fanfare, that introduced crops and livestock account for 98 per cent of all food produced in the US, with a value of $800 billion per year. So, however much we discover (or imagine) aliens are costing us, it's going to be small relative to that. When adding up the costs of introduced species, it's traditional to ignore anything on the positive side of the balance sheet, so just make a mental note of that and move on.

We then come to a few generalities, starting with Wilcove et al. (1998), always the first port of call for those in need of support for the view that alien Armageddon is just around the corner (I have already briefly discussed the failings of this source in Chapter 6; for a more detailed critique, read Gurevitch and Padilla, 2004, or chapter 11 in Mark Davis's book *Invasion Biology*). Then comes the curious statement that 'In other regions of the world, as many as 80 per cent of the endangered species are threatened and at risk due to the pressures of non-native species.' The source for this remarkable claim is a brief news item in the magazine *New Scientist*, reporting a modelling study showing that alien plants may be reducing Cape Town's water supplies. Whatever the truth of that particular claim, the single relevant sentence is 'Of the 1406 fynbos species listed as endangered in South Africa's Red Data Book, some 80 per cent have been put there by the invading aliens, say scientists'. So 'other regions of the world' means a small corner of South Africa, and the 80 per cent claim comes from unidentified 'scientists'. Given the track record of such claims elsewhere, such second-hand evidence has to be treated with scepticism.

We then come to the assertion that alien species are costing the USA $100–200 billion per year. A glance at the table that accompanies this sum quickly reveals that at least half of this is attributable to the massive costs of human diseases, and nearly all of *that* is influenza. Recent flu pandemics may well have originated in eastern Asia, but it seems a stretch of the imagination to call flu an invasive alien species, if only because efforts to control flu are radically different in nature from those prompted by what we normally understand as 'invasive species'.

When we get to specific costs, under the subheading 'plants', there's rather a lot about our old friend purple loose-strife, relying mainly on the suspect sources I examined in Chapter 4. We are then told that purple loosestrife costs $45 million per year in 'control costs and forage losses'. The source for this gem is the National Wildlife Refuge Associa-tion website, which reports, out of nowhere, that loosestrife costs '$45 million for habitat restoration and control methods'. Now forgive me for an outbreak of pedantry at this point, but 'forage losses' does at least imply that loosestrife actually costs something (in addition to the costs of trying to control it), but 'habitat restoration and control methods' does not. In fact the National Wildlife Refuge Association have it right: the 'costs' of loosestrife amount to the expense of 'control' (presumably spraying and mechanical control), followed by the 'habitat restoration' needed to repair the damage caused by the control. Since we saw in Chapter 4 that there is no evidence that loose-strife actually causes any harm at all, one can't help thinking that just letting it alone would be better for all concerned. Not to mention cheaper.

Moving on, we learn that two invasive alien vertebrates are dogs and horses. Feral dogs cost $10 million per year in livestock losses, and $250 million per year in the direct and indirect costs of dog bites. Feral horses cost $5 million per year in forage losses. Now excuse me for being picky again, but calling dogs alien doesn't sound quite right. Domestic dogs were brought to the USA by Native Americans, which means they've been around an awfully long time. Not only that, dogs are only domesticated wolves, and wolves are US natives. And calling the horse alien is simply wrong; it is, as I pointed out earlier, a formerly extinct (but now reintroduced) native. But quite apart from the question of the nativeness or otherwise of dogs and horses, we are again in the familiar business of ignoring the positive half of the balance sheet. The subject is never mentioned, but presumably we are supposed to infer that the social and economic value of dogs and horses is zero.

Fish don't get much of a mention, except that economic losses due to exotic fish are 'conservatively' put at $5.4 billion per year. There's no way of telling how accurate this figure is, since it is based on 'unpublished data'. We should bear in mind, however, that sport fishing is very big business indeed in the US, valued at $69 billion per year, and that many important sport fish are introduced. For example the South American peacock bass is one of America's most popular sportfish, worth more than $8 million a year in Florida alone.

Under insects, the star of the show is the South American fire ant, reputed to be costing the US $2 billion per year, although as usual much of this is not the cost of actual damage caused by the ant but the costs of control. But how much of a problem is the fire ant? Here you don't need my opinion at all, because here's what Rachel Carson had to say in *Silent Spring*, one of the very few books that can be said to have changed the course of history:

> During most of the forty-odd years since its arrival in the United States the fire ant seems to have attracted little attention. The states where it was most abundant considered it a nuisance, chiefly because it builds large nests or mounds a foot or more high. These may hamper the operation of farm machinery. But only two states listed it among their twenty most important insects pests, and these placed it near the bottom of the list. No official or private concern seems to have been felt about the fire ant as a menace to crops or livestock.

> With the development of chemicals with broad lethal powers, there came a sudden change in the official attitude towards the fire ant. In 1957 the United States Department of Agriculture launched one of the most remarkable publicity campaigns in its history. The fire ant suddenly became the target of a barrage of government releases, motion pictures, and government-inspired stories portraying it as a despoiler of southern agriculture and a killer of birds, livestock and man. A mighty campaign was announced, in which the federal government in cooperation with the afflicted states would ultimately treat some 20,000,000 acres in nine southern states.

Carson then spends some time carefully examining the claims that the fire ant is a serious threat to agriculture, wildlife and human health, dismissing all of them. *Silent Spring* is by its very nature a passionate, polemical book, but Carson reserves some of her most vitriolic invective for the blanket spraying of dieldrin and heptachlor against the fire ant:

> Never has any pesticide programme been so thoroughly and deservedly damned by practically everyone except the

beneficiaries of this 'sales bonanza'. It is an outstanding example of an ill-conceived, badly-executed, and thoroughly detrimental experiment in the mass control of insects, an experiment so expensive in dollars, in destruction of animal life, and in loss of public confidence in the Agriculture Department that it is incomprehensible that any funds should still be devoted to it.

A final irony is that one of the principal sins laid at the door of the fire ant by the US chapter of *Biological Invasions* is a decline in the bobwhite quail, but, as Carson reported, the highly toxic organochlorine pesticides used against the fire ant all but eliminated quail from the treated areas. There's no doubt fire ants can have negative effects on quail, but it's a huge leap to infer that the ants are responsible for the bird's long-term decline. In fact, as usual, an alien species is taking

A fire ant paddles away from the scene of the crime.

the rap for a deeper problem. Here's what Reggie Thackston, a senior wildlife biologist with the Georgia Department of Natural Resources and coordinator of the Bobwhite Quail Initiative, has to say on the subject:

We've had a steady decline in the [quail] population since the '60s, with some natural fluctuations. While the population has been on a downward spiral, excellent weather conditions can sometimes result in increased numbers of quail for a short period of time. Good weather can't compensate over the long haul, however, for a loss of habitat, which began resulting in a decline in quail numbers as early as the 1920s. This decline reached a critical stage in the 1960s.

Habitat is the key. There is no mystery here about quail decline. Some might attribute it to fire ants or acid rain, but the bottom line is that in places where you have substantial acres of good habitat you will have quail. You can see that in Georgia. You can also look to the west in Texas, Oklahoma, and Kansas, where by accident they still have good blocks of good habitat and still have quail. It's all about land use changes.

You'll also not be surprised to learn that the fire ant programme that so incensed Rachel Carson was, of course, a complete, abject failure, with most states reporting more ants after the programme than before. In Rachel Carson's words, 'The most expensive, the most damaging, and the least effective programme of all.'

Finally, returning to the costs of alien species in the US, more than half a page on amphibians and reptiles mentions no amphibians and only one reptile: our old friend the brown tree snake. But, you are already thinking, I didn't know the brown tree snake had been introduced to the US. No, indeed, all the costs listed (which are without a doubt large and genuine) refer to Guam, which is an 'organised unincorporated territory of the United States'. Still, if you are keen on inflating the costs of species introduced to the US by counting everything including the kitchen sink, there's nothing to stop you including a small island on the far side of the Pacific. Interestingly, elsewhere in the book, the brown tree snake appears again (twice) in a chapter on invasive vertebrates of South Africa (somewhere else it doesn't occur).

In short, this account of the costs of aliens in the US is so badly drafted that it would be hard not to laugh if the estimates it contains were not taken so seriously; it is, for example, a key reference on the US Department of Agriculture website. But the real lesson is not any particular instance of incompetence or chicanery, but what it tells us about the mindset of those who have allowed themselves to be swept along by the 'invasions industry'. Essentially it's obligatory to accept the notion that alien species have only negative effects. Thus the 'costs' of alien species have a rather peculiar interpretation for those of us used to putting together the credit and debit sides of a balance sheet to arrive at a *net* cost. Not only that, but the 'costs' themselves need not include anything that actually leaves you worse off as a result of the existence or activities of a particular species. They can also include (or, indeed, consist entirely of) the costs of attempting to control the species, plus the costs of clearing up the mess left by such attempts, even if there's little evidence of any original harm (e.g. purple loosestrife or spotted knapweed), or if such attempts are never likely to succeed without tackling a deeper underlying problem (e.g. tamarisk). Of course these costs are real, in the sense that they represent actual dollars spent – the question is whether it was money worth spending.

The final lesson is that, whenever you hear eye-wateringly colossal sums mentioned in the context of costs of alien species, you should always enquire exactly what these costs are supposed to include, and where they came from.

#3 ALIENS ARE ALWAYS TO BLAME

Many people seem prepared to believe that any evidence of harm caused by an alien species, however outdated, second-hand, anecdotal, unreliable or discredited, is probably true. There are plenty examples of this phenomenon, but here are just a few.

The Nile perch is an iconic invasive species, almost up there with the brown tree snake. Its introduction to Lake Victoria in Africa is widely assumed to have been the direct cause of the extinction of many of the extremely diverse cichlid fish that formerly inhabited the lake. Since cichlids are small, and the Nile perch is big and predatory, this looks like an open-and-shut case. But closer inspection again reveals that things aren't quite so simple. Expansion of the Nile perch population in Lake Victoria led to the development of a major fishery, providing employment for local people and millions of dollars in foreign exchange from fish exports. But this success brought problems, including increased fishing pressure and major development around the lake, leading in turn to pollution from agricultural, sewage and industrial waste, and topsoil erosion and silting from deforestation of the lake margins (both to provide domestic fuel and to smoke fish). In the end it's very hard to separate the impact of this environmental degradation on the native cichlids from that of the Nile perch itself.

Another popular demon is American mink, which were imported to Britain to be farmed for their fur, but escaped on numerous occasions and established wild populations in the 1950s, which spread throughout the following decades. Establishment of this alien predator had negative effects on many native species, especially water voles and ground-nesting birds. Around the time the wild mink population was rapidly expanding (1960–1980), the native otter simultaneously declined, and it was widely believed (even in scientific circles) that the mink may have contributed to the decline

American mink – not guilty. And no match for British otters.

by competing for food. In fact, mink and otter diets overlap
very little, and it is now known that organochlorine pesticide
pollution was responsible for the otter decline.

Following the banning of organochlorines, a general
increase in water quality of British rivers, and rigorous protection, otters have recently increased in numbers and range
in Britain. A study in north-east England has convincingly
demonstrated that the establishment of otters in a catchment
leads relatively quickly to the loss of any mink in the neigh-
bourhood. Since it's hardly ever possible actually to observe
interactions between the two species, the mechanism isn't
known for sure, but again competition for food is unlikely.
Direct aggressive encounters seem most likely, with the mink
generally finding itself on the losing side; otters are three times
longer and ten times heavier than mink. In fact, the original
otter decline led to a form of mesopredator release very like

that involving dingoes and foxes in Australia. In Ireland, where there has always been a large and thriving otter population, mink were never anything like as successful as in England. The satisfying thing about this story is not just that it seems to be on its way to a happy ending (for the otter anyway), but that it's a nice example of an invasive species problem being solved by fixing the underlying environmental problem.

Sometimes a whole group of species comes in for unjustified criticism. Most people are fond of most birds, most of the time, so birds generally have escaped the extreme opprobrium heaped on invasive snakes, toads and rats. Recently, however, researchers have assembled 'evidence' to show that several birds introduced to Europe are so bad that a major effort should be made to eradicate them. Birds singled out for the firing squad include the Canada goose, ring-necked parakeet and monk parakeet, but near the top of the hit list is the sacred ibis. This African bird, introduced to France, Spain and Italy, is a generalist predator that feeds on fish, frogs, small mammals, reptiles, smaller birds and invertebrates, and it's for this behaviour that it finds itself in the dock. Specifically it is accused of being the driving force behind the decline of many species, and of preying on species that are endemic or listed as vulnerable, endangered or critically endangered by the IUCN (International Union for Conservation of Nature), leading to the local extinction of such species.

But there is no evidence for this claim; the only two sources that are supposed to support this view say 'the cases outlined above are believed to have had no serious impact on the populations of the species [birds] preyed upon'. Moreover, all species thought to suffer from predation by sacred ibis are designated as 'Least Concern' by the IUCN in Europe. Although the researchers 'observed predation of sacred ibises on newts may have detrimental effects on discrete populations of these endangered amphibians', again no evidence is offered to support their claim. Sacred ibis are also supposed to compete with cattle egrets and little egrets for nesting sites (you can find this claim on Wikipedia), but, since populations of both these species are currently increasing in Europe, the ibis can't be doing them too much harm. Wikipedia also asserts that 'these large predators can devastate breeding colonies of species such as terns', but a review of the predation of seabirds by ciconiiform birds (storks, herons, ibis and similar birds) reports only that 'The literature indicates that predation of seabirds by ciconiiform birds is limited to terns and is generally of little impact', which isn't quite the same thing.

Interestingly, the authors of the original paper responded to criticism of their exaggerated claims for the impacts of introduced birds as follows:

> We also do not agree with the claim that anecdotal and observational data should be ignored, whereas 'possible positive effects' should be taken into account, especially considering the fact that there are only few studies actually performed on impacts of invasive species. So it is more likely that the impacts are underestimated than overestimated, because of lack of studies.

Which seems to be saying that lack of evidence of *positive* effects of alien species means there aren't any, but lack of evidence of *negative* effects means things must be far worse than we suspected.

Many observers seem to be happy to believe that an increase in alien X must be the cause of a decline in native Y, if the increase and decline happen around the same time. More surprisingly, so strong is the urge to see cause and effect in almost any pair of vaguely correlated variables, that this willingness extends even to native species that are seen to be too aggressive – as we will see over page.

Sparrowhawks and magpies – caught in the act?

In recent decades, populations of many British garden songbirds have declined dramatically. The reasons for this decline are unclear, but for some gardeners the cause is only too obvious – the increase over the same period in abundance of the sparrow-hawk and the magpie, two major predators of songbirds. Since 1970, both the sparrowhawk and the magpie have spread eastwards in Britain, although the causes are very different. Sparrowhawks have spread into areas of intensive arable farming, following the decline in use of the organochlorine pesticides that had formerly eliminated them from these areas. Magpies, on the other hand, seem to have benefited from a fall in the numbers of lowland gamekeepers. It's possible to track the expansion of both birds, and the decline of songbirds, using the Common Birds Census (CBC), which has monitored numbers of birds at around 200 woodland and farmland sites since 1962.

The CBC can be used to test the hypothesis that songbirds declined at these sites when sparrowhawks or magpies, formerly absent, arrived from the west. Of the 23 songbirds examined, most have declined, but some, such as the chaffinch and blue tit, have increased. All 23 songbird species are taken as prey by sparrowhawks, while eggs or nestlings of many (but not all) are eaten by magpies.

First, the most obvious question: were there fewer songbirds at a site *after* either the sparrowhawk or magpie was first reported than there were before? The answer is clear: for most of the songbirds studied, a significant decline was correlated with the arrival of one or other of these two predators. It looks like an open-and-shut case, but like the otter and mink, it isn't. There's a risk that the decline of many songbirds during the 1970s and 1980s and the increase in their predators just happened to coincide by accident, and the latter was not actually the cause of

Not a pretty sight: a sparrowhawk takes a young garden bird. But we shouldn't jump to conclusions ...

the former. It's easy to see how this could happen by looking at the collared dove, which also underwent a remarkable population explosion around the same time, although this time from the east. It turns out that collared doves provide an even better *statistical* explanation for the decline of songbirds than sparrowhawks and magpies. Except that this time we know the expansion of the collared dove was not the cause of the decline of songbirds, since collared doves are not predators of songbirds or of anything else.

If, instead of just asking whether songbirds declined after their predators arrived, we ask if they declined *more* than they did everywhere else in the country (irrespective of the presence of sparrowhawks or magpies), all connection between songbirds and sparrowhawks or magpies disappears. The songbird decline definitely coincided with the expansion of their predators, but it happened everywhere at about the same time, *whether or not their predators were present*. The finger of blame for songbird decline, and for many other undesirable changes in the British countryside over the same period, almost certainly points at agricultural intensification. Sparrowhawks and magpies were just innocent bystanders.

In fact, the latest research reveals there is a generally positive relationship between songbirds and nest predators, i.e. more magpies and crows, more songbirds. Whatever you make of that surprising result, as the authors of the work say, it 'largely exonerates these predators as driving declines in passerine [songbird] numbers'.

It's not only animals that are accused of crimes they didn't commit – here's a plant example. Oak savannah in south-western British Columbia, Canada, is naturally very species-rich, but the surviving remnants are heavily invaded (and now dominated) by two European grass invaders: *Poa pratensis* and *Dactylis glomerata*. No native species has been lost, or at least only four species since the 1840s, losses which are easily accounted for by habitat destruction (more than 95 per cent of oak savannah has been lost), but many native species are now very rare or threatened. The understorey of savanna remnants is more than 80 per cent *Poa* and *Dactylis*, and it seems obvious that these two are responsible for the decline of the native plants. But careful experiments, involving mowing or weeding the dominant grasses and sowing seeds of native species, tell a different story. The conclusion is that the present grass-dominated state of these savannahs is a result of a relatively recent policy of fire suppression. Historical records show that unburned areas of savannah have always been dominated by perennial grasses, and although in the past these would have been natives, the two aliens are merely occupying a niche that would previously have contained native grasses. In short, the alien grasses are not the cause of the system's present low diversity, but are merely 'passengers' of changes caused by human interference.

#4 ALIENS ARE OUT TO GET US

Why are we quick to blame invasive species for anything and everything that appears to be going wrong, yet very slow to blame overfishing, pollution, habitat loss, overgrazing, climate change, intensive agriculture and all the other much more fundamental causes of our environmental problems? It's tempting to say that the real problems are just too big, too diffuse and basically too damned expensive. Tackling them in a way likely to make any difference would probably involve reductions in our standard of living, or at least changes in the way we live our lives. All that's true, but there are even deeper psychological reasons, connected to our evolutionary history and the way we assess risks. We find it hard to get worked up about abstract, impersonal threats – however big their potential damage – so malaria, climate change and obesity don't get anything like the attention they deserve. On the other hand, we are acutely sensitive to threats from clearly identifiable, discrete agents, so our anxiety about terrorists, violent criminals and paedophiles is out of all proportion to any harm we are likely to suffer at their hands.

Where do invasive species fit in? You might imagine they would belong with things that we have trouble taking seriously, like influenza and tuberculosis, but you'd be wrong. The popular media, and even scientists to some extent, have succeeded in personalising the threat from invasive species by describing them as a 'menace' or a 'plague', and crucially even attributing malicious intentions to them: 'evil aliens'. Purple loosestrife alone has been described in newspaper reports as an invader, menace, pest, plague, killer, scourge, monster, public enemy number one, enemy of wildlife, time bomb, disaster, nightmare, rogue, strangler, barbarian and, my personal favourite, the 'Freddy Krueger of plants'. Even the relatively neutral word 'alien', used by scientists in a value-free way to define non-native species, raises troubling political and moral issues when used casually, and an unthinking nativism may have roots in xenophobic and racist attitudes, or at least reinforce such attitudes if employed carelessly.

In the UK the *Independent* (really a rather sober and responsible newspaper) ran a story with the headline 'British creatures fight for survival in rivers and meadows as aliens stage invasion of the wild' above a story that begins 'An invasion of aliens is under way, threatening our native species with death and disease. Rogue animals from all corners of the world have colonised Britain, multiplying with ease despite often hostile environments which are quite different from those of their homelands.'

It's interesting to note that the UK Black Environment Network, whose mission is 'to enable full ethnic participation in the environmental and heritage sectors', is sufficiently concerned about the rather fine line between stigmatising alien species and doing the same to alien *humans* to campaign actively for a more tolerant attitude to both. Indeed it's hard to read about attempts to distinguish between effectively identical frogs or beavers (see Chapter 5) without being reminded of similar attempts to separate humans by skin colour in apartheid South Africa or by 'Jewishness' in Nazi Germany. Hitler, it might be recalled, was an advocate of nativism, believing strongly that gardens should be planted only with native species.

#5 ALIENS ARE BAD, NATIVES GOOD

The mantra of 'aliens bad, natives good' has tended to conceal the inconvenient truth that native species, especially when given an (often unintentional) opportunity by various forms of human interference, can be pretty badly behaved, too.

Michael Carey and colleagues have documented the example of the problems facing Pacific salmon in the American Pacific North-West. The Columbia River basin has been heavily modified by dam construction, waterway dredging, flood control measures and construction of artificial river channels, all tending to provide opportunities for a variety of native salmon predators.

The excellent nesting habitat provided by artificial islands, absence of native predators and abundant food supplies, including juvenile salmon from hatcheries, has allowed the population of Caspian terns to double since 1980. The terns now consume up to 15 per cent of juvenile salmon in the Columbia River estuary. The same dams and reservoirs have led to a massive expansion in the population of the northern pikeminnow, a major predator of young salmon. Meanwhile, salmon congregating near the fish ladders below the Bonneville Dam have proved irresistible to Pacific harbour seals, California sea lions and Steller sea lions, which now swim the 235 km inland from the estuary to partake of the bounty we have provided for them; they ate nearly 5,000 salmon in 2009.

The response to salmon predation by terns, pikemin-nows and marine mammals (all natives) displays all the usual hallmarks of treating the symptoms rather than the disease: attempting to relocate tern colonies, deter marine mammals by physical barriers, acoustic deterrents, pyrotechnics, harassment and relocation, and the Pikeminnow Sport-Reward Program, which rewards anglers for catching pikeminnows. These efforts have varied from moderately effective to completely useless, but all tend to be expensive, and none tackles the underlying problem of the dams and impoundments that caused the problem in the first place.

All the predictable difficulties of trying to manage the natural world are here compounded by societal attitudes.

Catch a Squawfish & Earn some Cash!

January 27, 2011 • 8 Comments

[f Like] 7 people like this. Be the first of your friends.

American anglers are recruited to eliminate a pesky native.

Having been trained to believe that only aliens cause this kind of mess, the public find it hard to understand why natives can be a problem. Pikeminnow control is fine with the public, but harassment and possible lethal control of charismatic marine mammals generally isn't (a court case brought by the Humane Society of the United States and the Wild Fish Conservancy prevented 'lethal removal' of sea lions). Public attitudes to terns are mixed, but there are worse problems: just like Caspian terns, double-crested cormorants (another native) have benefited from human habitat modification, increased in numbers and now threaten juvenile salmon. Meanwhile, cormorant colonies along the Pacific coast are declining, creating the perfect cormorant conundrum: a species that is declining in one part of its native range, but becoming a bit of a pest in another part. It doesn't help that state agencies like the Washington Department of Fish and Wildlife are supposed to protect native species *and* maintain recreational fisheries for popular sport

212

fish, aims that are not obviously compatible. The truth is that, as is usually the case, it is our modification of the habitat that is the root of the problem, but somehow it all seems more complicated when all the species involved are native. As Carey drily comments, 'Policy makers struggle to find acceptable solutions for contending with native invaders.'

And in case you think I've picked here on an isolated, unrepresentative example, it's worth noting that I had plenty to choose from. For example, the remarkable spread of western juniper (*Juniperus occidentalis*) across 9 million hectares of the Intermountain West in the USA since the late nineteenth century, largely owing to the huge decrease in frequency of fires since human settlement. Or the expansion of the American black vulture (*Coragyps atratus*), now up to 20 million birds, owing to its fondness for human rubbish, combined with its complete indifference to the presence of actual humans.

Mindful of the damage likely to be inflicted by too many badly behaved natives on the ongoing 'native good, alien bad' project, one of its chief defenders, Daniel Simberloff, 'analysed the data' in an article for *Ecology* to show that natives are, almost without exception, a polite, well-mannered bunch. But the data can show whatever we want them to show, provided we choose them right. In this case the data, focusing on invasive plants in the USA, came from a systematic search of the scientific literature for reports of 'invasive aliens' and 'invasive natives'. But this was a search deeply flawed from the start, because among the invasion biology community, 'invasive' and 'alien' go together like 'bread' and 'butter'. Simply put, natives are unlikely to be reported as invasive. But, Simberloff protests, we thought of that, so we searched not just for natives described as 'invasive', but also for those that were 'encroaching' or 'expanding', which might be more likely to pick up studies of natives. But you can refine the search until you're blue in the face – in the end, you can pick up only what is reported, and natives just aren't news to anything like the extent aliens are. 'Invasive aliens' generate not only column inches (in the popular media and in scientific journals), but crucially also they grab the attention of funding agencies. 'Expanding' or 'encroaching' natives are less likely to be noticed, and even less likely to be reported.

But even if you wanted to report those rampaging natives, there's always the old *Ghostbusters* question: who ya gonna call? A major source (probably *the* major source) of raw data for any analysis of published work on invasions is the journal *Invasion Biology* (editor-in-chief D. Simberloff), which publishes several hundred papers per year, virtually all of them about introduced species. For example, of the 247 papers published in 2011, as far as I can tell only one was a study of an invasive native. There's no particular prohibition of papers about natives, but it's simply understood among contributors to *Invasion Biology* that 'invasive' means 'invasive alien'.

The Pacific salmon example with which I began this section is instructive. A few more cormorants here, a few less there, or a few sea lions turning up in the wrong place, are typical examples of 'invasive natives'. That is, they are not seen as interesting in themselves, not really worth reporting; they coalesce into an interesting story only because they combine to impact on something we *are* interested in: salmon. If there's no salmon – and there usually isn't – invasive natives tend not to appear on our radar.

CHAPTER TWELVE

WHERE DO WE GO FROM HERE?

In 2011 I contributed to a comment in the journal *Nature* entitled 'Don't judge species on their origins'. We made the modest assertion that 'Nativeness is not a sign of evolutionary fitness or of a species having positive effects ... and classifying biota according to their adherence to cultural standards of belonging, citizenship, fair play and morality does not advance our understanding of ecology'. We went on to assert that our attitude to alien species seems rooted in a mythical past where foreign species invaded pristine native habitats, but in a world completely transformed by humans, *all* species now find themselves effectively strangers; and the hope that the native species that were happiest before that transformation will continue to be so seems likely to be disappointed. We concluded: 'We are not suggesting that conservationists abandon their efforts to mitigate serious problems caused by some introduced species, or that governments should stop trying to prevent potentially harmful species from entering their countries. But we urge conservationists and land managers to organize priorities around whether species are producing benefits or harm to biodiversity, human health, ecological services and economies.' In other words, treat species on their merits, where those merits are defined by the best available scientific evidence.

A swift rejoinder came back in a letter with a remarkable 141 signatories (led by the tireless D. Simberloff) entitled simply 'Non-natives: 141 scientists object'. This very brief letter accused us of attacking two straw men. Specifically, it said:

> First, most conservation biologists and ecologists do not
> oppose non-native species per se – only those targeted
> by the Convention on Biological Diversity as threatening
> 'ecosystems, habitats or species'. There is no campaign
> against all introductions ... Second, invasion biologists and
> managers do not ignore the benefits of introduced species.

If we've learned anything in the course of this book, I hope it's that there is no truth in either of these claims. Invasion biologists and managers *do* almost routinely ignore the benefits of introduced species. And non-native status is widely seen as a 'mark of Cain' that means it's a wise precaution to assume that all aliens are up to no good unless conclusively proven otherwise. Indeed, Simberloff has written elsewhere that '*Every* proposed introduction must receive the scrutiny currently reserved for species known to have caused harm elsewhere' and 'every proposed introduction be viewed as potentially problematic until substantial research suggests otherwise'.

Note that since introduced species are only rarely subject to substantial research before sentence is passed, this amounts to a call for nearly all introduced species to be deemed undesirable for the foreseeable future. Also recall the quote in the previous chapter about alien birds, essentially taking the view that substantial research is hardly necessary for those species that we just 'know' are causing trouble; contrary to popular belief, no news is bad news. I accept that most conservation biologists and ecologists do not actively oppose *every* alien species, but I suspect the reason is primarily a matter of practicality rather than principle: there just aren't enough hours in the day. In a recent survey, more invasion biologists agreed than disagreed with the statement that 'Exotics are an unnatural, undesirable component of the biota and environment'.

Yet can it be possible that all these 141 scientists are part of some conspiracy or mass delusion? I don't think so. The problem, rather, lies in looking at the world in a way that makes the notion of alien species threatening ecosystems or species not a hypothesis but a truism. For example, invasion biologists worry a lot about biodiversity, but only native biodiversity; introduced species are not allowed to contribute to biodiversity. Thus alien species can never add to biodiversity, they can only reduce it, not because there's anything special about them, but because we've chosen to define biodiversity in a way that makes that inevitable.

Much the same applies to the concept of 'harming' ecosystems. As Mark Sagoff has pointed out, 'If one defines as "harm" any significant change a non-native species causes, the statement that non-native species harm ecosystems represents a tautology.' Thus, if the presence of zebra mussels means clearer water, more aquatic plants, more fish and more wildfowl, then that's 'harmful' by definition, presumably because the previous water clarity, plants, fish and wildfowl were just right. Of course, muddier water and fewer plants, fish and wildfowl would have been 'harmful', too. What we seem to have here is an appeal to a vague concept of ecosystem 'health' or 'integrity', where it's never quite clear what either term means, although it's implicit that alien species are inimical to both.

We can now see how the two sides in the above debate can both be sincere and, in a curious way, both right too. Once you fall down the rabbit-hole or step through the looking glass, you enter a world where you are indeed not ignoring the benefits of introduced species, because (by definition) there aren't any; where you are indeed only concerned with species that threaten native ecosystems or habitats, because all introduced species (by definition) pose such threats. Or at least we have to assume they do until every introduced species has been subject to the same level of attention as purple loosestrife, which is approximately never.

But we have now moved well outside the theory and practice of ecology as normally understood, and we are instead appealing to ecology to tell us what is natural and healthy. But ecology is a science, not an ideology, and these concepts are both value-laden and subjective. Ecology should not – cannot – act as an authority that promotes one vision of nature at the expense of others. Value-judgements lie outside the borders of science; if a wetland, formerly dominated by *Typha* (cattail or reedmace), is now dominated by purple loosestrife, no science can tell us that is a bad outcome, or a good one. By pretending that it can, we risk losing the one thing that is absolutely essential to the conservation of biodiversity, which is the informed consent of the public. It's been pointed out, although one gets the impression it hasn't really been grasped, that attempted eradication or control is a waste of time if active support (or at least passive cooperation) is required of people who believe the target species to be useful, desirable, attractive or simply harmless.

In the Galapagos example described in Chapter 8, the biggest single cause of failure was lack of cooperation from landowners. Since there was no public consultation, and the first thing most knew about the project was when they were asked for permission to enter their land, this wasn't exactly surprising. Landowners were simply unimpressed by the arguments, especially when the target species was (from their perspective) a valuable pasture species or timber tree.

The devil's claw eradication programme (again, see Chapter 8) has been widely praised for a number of positive outcomes. As the main report observed: 'Young rangers gained skills in leadership, project management, bush living, and the use of GIS. Eighty-two Aboriginal Rangers participated in the project .... [and] the surveys allowed them access to their country and to carry out traditional burning, hunting and maintain sacred sites. Furthermore, many non-aboriginal people got first-hand experience of working with Aboriginal Australians and gained some insight into how they manage land. Over 200 local, national and international volunteers participated, each gaining experience to share with the world.' No one would argue with the value of those outcomes, but how much better if they had grown out of an eradication programme that had (a) some chance of success and (b) targeted a species that actually deserved all that attention.

The not unreasonable tendency of film and television to focus on extraordinary nature in apparently pristine habitats already tends to endorse the impression that 'real' nature is found only in remote, special places, far from where most of us live. Promoting the idea that alien species somehow 'don't count', and are in fact some kind of living pollution can only increase what already threatens to become a serious disconnect between people and nature, and between people and scientists. Most of us live in cities, and our daily contact with nature, if we have one at all, is predominantly with introduced species. Even native species are effectively exotic in cities, in the sense that they have colonised an entirely artificial, human-transformed environment. The 'scientific' view that such species have no value or meaning is the exact opposite of most people's experience, as evidenced by the long history of contact with cultivated plants and domestic animals.

The conservation argument has to make sense to the public, and a rigid separation of indigenous and exotic does not make sense. In fact, the public are ahead of the scientists in this respect; they intuitively grasp that in the world they inhabit every day, some species (some alien, some native) are clearly well adapted, while others are not, and they frequently neither know nor care that some of the well-adapted species don't 'belong'. As one invasion biologist put it (without irony): 'the public does not readily distinguish between native and non-native species: as long as an animal looks nice and is not threatening people or causing undue harm, the public tends to view species equally'. Quite – you just can't trust the public to get anything right, can you? Not having been educated to venerate the idea of nativeness, they are just as likely to value novelty, beauty, diversity or rarity. In fact, of course, so does the conservation community, privileging the native and denigrating the exotic only when it suits them.

Some people like purple loosestrife, or sacred ibis, or ringed-neck parakeets, or starlings, and even *Rhododendron ponticum*, and who is to say they are wrong?

〜 〜 〜

Previous experience tells me that there are people who, for one reason or another, will misunderstand the message of this book. So, just to make absolutely sure we understand each other, am I suggesting we should stop trying to slow the spread of alien species, or trying to control or eradicate the small minority of species that do cause serious problems? No, I'm not. Would Guam be better off without the brown tree snake, Australia without the cane toad, and several Pacific islands without the predatory snail *Euglandina*? Yes they would, although it's worth

Cane toads gathering below a water pipe in Queensland – Australia would be better off without them.

noting that the last two are deliberate introductions intended to control previous, less damaging introductions.

I *am* saying that we should commence any attempt to control alien species with our eyes wide open. We should be certain, from an honest, objective analysis of the best available evidence of its positive and negative impacts, that our intended target is causing *net* harm. We should be sure that the alien species itself is the problem, and not merely a symptom of an underlying environmental problem. Nor should we tinker with our definitions of 'native' and 'alien' to suit our prejudices. When we come to try to add up the economic and (especially) environmental costs of alien species, we should not define those costs in a circular, question-begging way, such that alien species are harmful by definition. And, although this should be too obvious to need saying, I'm going to say it anyway: the costs of trying to control a species should not – logically, *cannot* – form any part of the justification for attempting that control in the first place.

We should also be reasonably confident that the benefits of control will outweigh the costs, bearing in mind that since eradication is rarely an option the costs of control may extend indefinitely into the future. If eradication really is the aim, we should recall that most eradication attempts, even when all the indicators are positive (which they rarely are), end in failure. Since many alien species thrive in disturbed, early-successional environments, we should consider the possibility that our control attempts themselves may create more of these conditions and thus make things worse. We should be mindful that many invasions go through an initial 'boom' phase, but eventually settle down at a lower level, not least because of adaptation by both the invader and the invaded ecosystem; accordingly, we should try not to react in panic to the early stages of an invasion. Last but not least, we should choose targets, methods and strategies that offer some prospect of success. While one can reasonably argue that things would have been worse if no control had been attempted, something is wrong when, after a 15-year programme costing nearly half a billion dollars, the area of South Africa occupied by introduced plants has actually increased.

Adopting the approach outlined above, which in reality is no more than a belated outbreak of common sense, will waste a lot less time and money on wild-goose chases. We will also, by crying wolf less often, run a much lower risk of alienating the public. There is a deep well of goodwill towards attempts to manage introduced species, but it's not bottomless, and we squander it on lost causes at our peril.

I also suggest, and as a professional ecologist I hate to have to say this, that we should stop expecting too much from ecology. I described the ability to predict in advance which species will become invasive as the Holy Grail of invasion biology, but such predictions and the Grail seem to share the unfortunate property of not actually existing. Looking for traits of invasive species is an interesting intellectual exercise that is likely to keep a few ecologists in employment for the foreseeable future, but all the evidence suggests it's unlikely to be of much practical use.

Finally, in a world in which the spread of alien species is only one small part (and far from the most important) of the complete transformation of the biosphere by human activity, we should stop thinking that we can turn the clock back to some pristine, pre-human golden age, even if we had any idea what that pristine state looked like. We should instead focus on getting the best out of our brave new invaded world.

ACKNOWLEDGEMENTS

This book could not have been written without the hard work of many hundreds of scientists, to all of whom I am deeply grateful. Even reading the work of those with whom I do not entirely agree has been helpful in sharpening – and occasionally even demolishing – some of my flabbier arguments. There is no way I can thank everyone individually, but most of the sources I consulted are cited in the Notes, following. The referencing is not as comprehensive as in an academic text (this is not a textbook), but there is enough, I hope, to allow a reader to discover my important sources of ideas and information.

Just as no battle plan survives contact with the enemy, few bright ideas survive their first encounter with the published, peer-reviewed literature. Normally such contact quickly reveals that the idea was certainly less original and often less bright than one had imagined, so it's worth remarking here on an unusual finding: that the more I read, the more I felt the convictions expressed in this book are right, and that much of what passes for invasion biology is poorly supported hype.

A few people deserve individual thanks. The work of David Pearman, former President of the Botanical Society of the British Isles, has been a constant source of inspiration, and I draw heavily on his painstaking work showing that our understanding of the 'native British flora' is not as good as most people think. Matt Chew, of Arizona State University, very kindly let me see a copy of his thought-provoking and extremely well-researched PhD thesis. I could not have written – even begun to have written – much of Chapter 2 without it. Mark Davis, of Macalester College, Minnesota, reignited my semi-dormant interest in invasion biology in the late 1990s, and since then has done more than anyone to keep my interest in the topic alive. In particular, Mark's 2009 book, *Invasion Biology*, has been an invaluable source of insight and ideas. Mark also very kindly agreed to read and comment on an earlier draft of this book. Matt, Mark and I were co-authors on our 2011 paper in *Nature*, 'Don't judge species on their origins', and all eighteen of my fellow authors deserve my sincere thanks, both for being willing to stick their heads above the parapet, and for being one of the major sources of inspiration for this book.

But of course, no book is much use until someone is willing to publish it, which brings me happily to Mark Ellingham of Profile Books. More than once, someone has shown up just when I was beginning to think I had written something that no one wanted to read, and my saviour on this occasion was Mark. He not only saw clearly what I was trying to say, despite my efforts to make that difficult, but was also able to explain, kindly but firmly, that taking too much for granted and thus starting the book in the middle was probably not a good idea. If this book is now any good, Mark can take much of the credit, and if it's not, well, if it had been left to me it could have been much worse. Sincere thanks also to Henry Iles (design), Nikky Twyman (proofreading) and Diana LeCore (indexing).

Thanks, as ever, to Pat for her encouragement during the writing of this book, including a willingness to defer the hoovering at crucial moments. I know, Rowan, that when you read this you will be my most loyal supporter, as usual. And Lewis, what can I say, except that I still haven't given up on writing a book you would consider reading.

PHOTO CREDITS

NOTES

I have benefited from countless books, papers, reports and conversations related to this topic over the years. Rather than a comprehensive bibliography, this section provides sources for specific information and arguments presented in the text.

INTRODUCTION

You can read about the history of the camel family (*Camelidae*) in almost any biogeography textbook, or indeed Wikipedia.

Although long periods of continuous biotic and environmental change can be studied from ocean sediment cores, it's rare to find very long continuous records on land. The information about the Bogotá basin in the tropical high Andes of Colombia comes from V. Torres et al. (2013) 'Astronomical tuning of long pollen records reveals the dynamic history of montane biomes and lake levels in the tropical high Andes during the Quaternary', *Quaternary Science Reviews*, 63, 59–72. The pollen grains preserved in this sediment column tell us what the vegetation was like at every moment during the last two million years.

CHAPTER 1

For background information on plate tectonics, the history of Gondwana and the northern supercontinent (Laurasia), and the Great American Interchange, Wikipedia is as good as anywhere. For the routes by which various species recolonised northern Europe at the end of the last ice age, see G.M. Hewitt (1999) 'Post-glacial re-colonization of European biota', *Biological Journal of the Linnean Society*, 68, 87–112.

Dispersal of seeds by sheep during traditional transhumance in Spain is described in P. Manzano and J.E. Malo (2006) 'Extreme long-distance seed dispersal via sheep', *Frontiers in Ecology and the Environment*, 4, 244–248. Dispersal of Metrosideros to Hawaii' is in S.D. Wright et al. (2001) 'Stepping stones to Hawaii: a trans-equatorial dispersal pathway for Metrosideros (Myrtaceae) inferred from nrDNA (ITS + ETS)', *Journal of*

Biogeography, 28, 769–774. Dispersal of Amphisbaenians, or worm lizards, by rafting across the Atlantic is reported by N.Vidal et al. (2008) 'Origin of tropical American burrowing reptiles by transatlantic rafting', *Biology Letters*, 4, 115–118.

Roman plant introductions to Britain are listed in M. van der Veen, A. Livarda and A. Hill (2008) 'New plant foods in Roman Britain – dispersal and social access', *Environmental Archaeology*, 13, 11–36. For vertebrates introduced to the USA, see G.W. Witmer et al. (2007) 'Management of invasive vertebrates in the United States: an overview', in *Managing Vertebrate Invasive Species: Proceedings of an International Symposium* (ed. G.W.Witmer,W.C. Pitt and K.A. Fagerstone), Fort Collins, CO: USDA/ APHIS Wildlife Services, National Wildlife Research Center.

For an account of the 'tens rule' which suggests that around 10 per cent of all introduced plants go on to escape into the wild, 10 per cent of these become genuinely naturalised, and only 10 per cent of this number go on to become pests, see Mark Williamson's 1996 book *Biological Invasions* (London: Chapman and Hall). The curious facts about the snails of Krakatau and the modern location of Plymouth Rock are from Steve Jones' 1999 book *Almost Like a Whale* (London: Doubleday).

For a glimpse of just how unlike today the Earth has been for most of its history, and how prone to (often poorly understood) catastrophes, see David Beerling's 2007 book *The Emerald Planet,* Oxford: Oxford University Press. For much the greater part of the history of life, the Earth has been warmer than today, often much warmer. It may be hard to believe that breadfruit, now native to tropical south-east Asia, once grew on the shores of Greenland, but the fossil evidence is clear that it did.

CHAPTER 2

Much of Chapter 2 is derived from Matt Chew's excellent 2006 Arizona State University PhD thesis 'Ending with Elton; preludes to invasion biology'. Matt's thesis was also useful for some parts of Chapter 1. An abridged version of many of the key ideas in Matt's thesis can be found in M.K. Chew and A.L. Hamilton (2010) 'The rise and fall of biotic nativeness: a historical perspective', in *Fifty Years of Invasion Ecology: The Legacy of Charles Elton* (ed. D.M. Richardson), Chichester: Wiley-Blackwell, pp. 35–47.

The quote about our differing ethical responsibilities to native and alien species is from J.C. Russell (2012) 'Do invasive species cause damage? Yes', *Bioscience,* 62, 217. Derek Ratcliffe's 1977 *Nature Conservation Review* is published by Cambridge University Press and is still in print. The David Shukman quote is from his 2011 book *An Iceberg as Big as Manhattan* (London: Profile Books), and the quote from 'nine American invasion biologists' is from M.W. Fall et al. (2011) 'Rodents and other vertebrate invaders in the United States', in *Biological Invasions: Economic and Environmental Costs of Alien Plant, Animal, and Microbe Species* (ed. D. Pimentel), Boca Raton, FL: CRC Press, pp. 381–410.

The quote about the relative costs of invasive species and natural disasters is from A. Ricciardi, M.E. Palmer and N.D. Yan (2011) 'Should biological invasions be managed as natural disasters?' *Bioscience*, 61, 312–317. My concluding comment about impending global shortages of water is based on C.J. Vörösmarty et al. (2000) 'Global water resources: vulnerability from climate change and population growth', *Science*, 289, 284–288.

CHAPTER 3

A short account of the brown tree snake on Guam is S.L. Pimm (1987) 'The snake that ate Guam', *Trends in Ecology & Evolution*, 2, 293–295, while a longer account can be found in G.H. Rodda and J.A. Savidge (2007) 'Biology and impacts of pacific island invasive species. 2. *Boiga irregularis*, the Brown Tree Snake (*Reptilia: Colubiridae*)', *Pacific Science*, 61, 307–324.

Information about the zebra mussel, and particularly about its negative impact on native freshwater mussels, is from A. Ricciardi, R.J. Neves and J.B. Rasmussen (1998) 'Impending extinctions of North American freshwater mussels (Unionoida) following the zebra mussel (*Dreissena polymorpha*) invasion', *Journal of Animal Ecology*, 67, 613–619.

My account of the history of tamarisk in the USA is from Matt Chew's 2009 paper 'The monstering of tamarisk: how scientists made a plant into a problem', *Journal of the History of Biology*, 42, 231–266.

The account of purple loosestrife is largely from a special report by the United States Fish and Wildlife Service, by D.Q. Thompson, R.L. Stuckey and E.B. Thompson (1987) 'Spread, impact, and control of purple loosestrife (*Lythrum salicaria*) in North American wetlands', United States Department of the Interior, US Fish and Wildlife Service, Fish and Wildlife Research no. 2, Washington, DC, while the 'weed of the week' quote is from www.invasive.org/weedcd/pdfs/wow/purple-loosestrife.pdf

CHAPTER 4

The opening paragraph of this chapter is based on H.A. Hager and K.D. McCoy (1998) 'The implications of accepting untested hypotheses: a review of the effects of purple loosestrife (*Lythrum salicaria*) in North America', *Biodiversity and Conservation*, 7, 1069–1079, while the Canadian study on the Bar River in Ontario is described in M.A. Treberg and B.C. Husband (1999) 'Relationship between the abundance of *Lythrum salicaria* (purple loosestrife) and plant species richness along the Bar River,

Canada', *Wetlands*, 19, 118–125. The 2010 review of the impact of purple loosestrife is by Claude Lavoie: 'Should we care about purple loosestrife? The history of an invasive plant in North America', *Biological Invasions*, 12, 1967–1999. The research showing that phenol-rich foliage is common to purple loosestrife and its close North American relatives is J.S. Cohen, J.C. Maerz and B. Blossey (2012) 'Traits, not origin, explain impacts of plants on larval amphibians', *Ecological Applications*, 22, 218–228.

The report of genetic changes in toxicity of garlic mustard in eastern North America is R.A. Lankau et al. (2009) 'Evolutionary limits ameliorate the negative impact of an invasive plant', *Proceedings of the National Academy of Sciences of the United States of America*, 106, 15362–15367.

The studies mentioned on the impacts of Himalayan balsam in Britain and mainland Europe are M. Hejda and P. Pysek (2006) 'What is the impact of *Impatiens glandulifera* on species diversity of invaded riparian vegetation?', *Biological Conservation*, 132, 143–152; P.E. Hulme and E.T. Bremner (2006) 'Assessing the impact of *Impatiens glandulifera* on riparian habitats: partitioning diversity components following species removal', *Journal of Applied Ecology*, 43, 43–50; I. Bartomeus, M. Vila and I. Steffan-Dewenter (2010) 'Combined effects of *Impatiens glandulifera* invasion and landscape structure on native plant pollination', *Journal of Ecology*, 98, 440–450; and C. Ammer et al. (2011) 'Does tree seedling growth and survival require weeding of Himalayan balsam (*Impatiens glandulifera*)?', *European Journal of Forest Research*, 130, 107–116.

The opening paragraphs of the account of tamarisk are from J.C. Stromberg et al. (2009) 'Changing perceptions of change: the role of scientists in *Tamarix* and river management', *Restoration Ecology*, 17, 177–186. The Invasipedia account of tamarisk can be found at http://wiki.bugwood.org/ Tamarix_spp, while the reference to Colorado water rights is taken from T.P. Barnett and D.W. Pierce (2009) 'Sustainable water deliveries from the Colorado River in a changing climate', *Proceedings of the National Academy of Sciences of the United States of America*, 106, 7334–7338.

J. Gurevitch and D.K. Padilla (2004) 'Are invasive species a major cause of extinctions?', *Trends in Ecology & Evolution*, 19, 470–474, asserts that there is a lot more to the decline of native unionid clams than any effect of zebra mussels, prompting a reply from A. Ricciardi (2004): 'Assessing species invasions as a cause of extinction', *Trends in Ecology & Evolution*, 19, 619, followed by a further response from Gurevitch and Padilla in 2004: 'Response to Ricciardi. Assessing species invasions as a cause of extinction', *Trends in Ecology & Evolution*, 19, 620.

The US Army Corps of Engineers Zebra Mussel Research Program can be found at http://el.erdc.usace.army.mil/ansrp/ANSIS/ansishelp.htm. The report of the potential use of zebra mussels in water filtration is P. Elliott,

D.C. Aldridge and G.D. Mogyridge (2008) 'Zebra mussel filtration and its potential uses in industrial water treatment', *Water Research*, 42, 1664–1674. An account of the impact of round gobies on zebra mussels can be found in A.M. Lederer et al. (2008) 'Impacts of the introduced round goby (*Apollonia melanostoma*) on Dreissenids (*Dreissena polymorpha and Dreissena bugensis*) and on macroinvertebrate community between 2003 and 2006 in the littoral zone of Green Bay, Lake Michigan', *Journal of Great Lakes Research*, 34, 690–697. The latest research suggests that the round goby could significantly reduce numbers of zebra mussels: R. Naddafi and L.G. Rudstam (2014) 'Predation on invasive zebra mussel, *Dreissena polymorpha*, by pumpkinseed sunfish, rusty crayfish, and round goby', *Hydrobiologia*, 721, 107–115.

The two papers about the improving abilities of native fish and crabs to eat zebra mussels are: N.O.L. Carlsson, O. Sarnelle, and D.L. Strayer (2009) 'Native predators and exotic prey – an acquired taste?', *Frontiers in Ecology and the Environment*, 7, 525–532; and N.O.L. Carlsson et al. (2011) 'Biotic resistance on the increase: native predators structure invasive zebra mussel populations', *Freshwater Biology*, 56, 1630–1637.

The idea that islands are highly susceptible to invasion is only one invasion hypothesis for which the evidence is not very good: J. Jeschke et al. (2012) 'Support for major hypotheses in invasion biology is uneven and declining', *NeoBiota*, 14, 1–20. I return to this question in Chapter 6.

CHAPTER 5

David Pearman has written two key papers on the 'doubtful' members of the British flora. C.D. Preston, D.A. Pearman and A.R. Hall (2004) 'Archaeophytes in Britain', *Botanical Journal of the Linnean Society*, 145, 257–294, looks at the archaeophytes (species introduced before 1500), while D.A. Pearman (2007) '"Far from any house" – assessing the status of doubtfully native species in the flora of the British Isles', *Watsonia*, 26, 271–290, looks at more-recent introductions. Plantlife's county flowers can be found at www.plantlife.org.uk/wild_plants/county_flowers/.

The evidence for (and against) the UK native status of the white-clawed crayfish can be found in D.M. Holdich, M. Palmer and P.J. Sibley (2009) 'The indigenous status of *Austropotamobius pallipes* (*Lereboullet*) in Britain', in *Crayfish Conservation in the British Isles, Conference Proceedings*, British Waterways Offices, Leeds, 25 March.

The report in the *Independent* newspaper about the 'unauthorised' Tay beavers can be found at www.independent.co.uk/environment/nature/murky-waters-why-are-beavers-being-sent-to-the-zoo-2164708.html. The quote about the dangers of introducing inbred beavers to the UK is from D.J Halley

(2011) 'Sourcing Eurasian beaver *Castor fiber* stock for reintroductions in Great Britain and Western Europe', *Mammal Review,* 41, 40–53.

The evidence that dingoes suppress cats and foxes is from C.N. Johnson, J.L. Isaac and D.O. Fisher (2007) 'Rarity of a top predator triggers continent-wide collapse of mammal prey: dingoes and marsupials in Australia', *Proceedings of the Royal Society B: Biological Sciences,* 274, 341–346. The dingo quote from the Melbourne *Herald Sun* can be found at www. heraldsun.com.au/news/victoria/dingoes-to-be-protected/story-e6frf7kx-1111117845024.

The convoluted story of Caribbean raccoons is told in K.M. Helgen and D.E. Wilson (2003) 'Taxonomic status and conservation relevance of the raccoons (*Procyon spp.*) of the West Indies', *Journal of Zoology,* 259, 69–76. The Don Wilson quote is from H. Nicholls (2007) 'Linnaeus at 300: the royal raccoon from Swedesboro', *Nature,* 446, 255–256.

The pool frog story is from T.J.C. Beebee et al. (2005) 'Neglected native or undesirable alien? Resolution of a conservation dilemma concerning the pool frog, *Rana lessonae*.' *Biodiversity and Conservation,* 14, 1607–1626, while the quote from Natural England can be found at www. naturalengland.org.uk/about_us/news/2013/290813.aspx.

CHAPTER 6

The definitions of niche and community are from the Cambridge International Examinations syllabus for AS and A level, and can be found at www.cie.org.uk/. The undergraduate biology textbook quoted is Neil A. Campbell, Jane B. Reece and Lawrence G. Mitchell (1999), *Biology* (5th edition), Menlo Park, CA: Benjamin Cummings. The 'previous book' referred to is K. Thompson (2010) *Do We Need Pandas? The Uncomfortable Truth About Biodiversity,* Dartington, UK: Green Books.

The quote about the relative performance of monocultures and mixtures of species is from B.J. Cardinale et al. (2006) 'Effects of biodiversity on the functioning of trophic groups and ecosystems', *Nature,* 443, 989–992. The American study showing that grassland diversity can be permanently increased by sowing extra species is D. Tilman (1997) 'Community invasibility, recruitment limitation, and grassland biodiversity', *Ecology,* 78, 81–92. The study showing overwhelmingly positive rather than compensatory dynamics in both plant and animal communities is J.E. Houlahan et al. (2007) 'Compensatory dynamics are rare in natural ecological communities', *Proceedings of the National Academy of Sciences of the United States of America,* 104, 3273–3277. The quote asserting that high diversity is a barrier to invasion is from T.A. Kennedy et al. (2002) 'Biodiversity as a barrier to ecological invasion', *Nature,* 417, 636–638.

The 1998 report that found alien species to be the second largest threat to the survival of natives in the USA is D.S. Wilcove et al. (1998) 'Quantifying threats to imperiled species in the United States', *Bioscience*, 48, 607–615. The data for acquisition of alien species and losses of native species by the USA come from T.J. Stohlgren et al. (2008) 'The myth of plant species saturation', *Ecology Letters*, 11, 313–322, while the similar data for oceanic islands come from D.F. Sax and S.D. Gaines (2008) 'Species invasions and extinction: the future of native biodiversity on islands', *Proceedings of the National Academy of Sciences of the United States of America*, 105, 11490–11497.

The Theory of Island Biogeography is a 1967 book by Edward O. Wilson and Robert MacArthur. The basic idea is reasonably well explained by Wikipedia (see http://en.wikipedia.org/wiki/Island_biogeography), while the effect of 'recreating Pangaea' by removing barriers to dispersal is explained by M.L. Rosenzweig (2001) 'The four questions: what does the introduction of exotic species do to diversity?', *Evolutionary Ecology Research*, 3, 361–367. The quote about the future of biodiversity in a more connected world is from M.A. Davis (2003) 'Biotic globalization: does competition from introduced species threaten biodiversity?', *Bioscience*, 53, 481–489.

The (very slow) ecology of the kauri tree is explained by J. Ogden, G.M. Wardle and M. Ahmed (1987) 'Population dynamics of the emergent conifer *Agathis australis* (*D Don*) *Lindl* (*Kauri*) in New Zealand. 2. Seedling population sizes and gap-phase regeneration', *New Zealand Journal of Botany*, 25, 231–242, while the recovery of alpine plants from the 'Little Ice Age' is documented in P.M. Kammer, C. Schob and P. Choler (2007) 'Increasing species richness on mountain summits: upward migration due to anthropogenic climate change or re-colonisation?', *Journal of Vegetation Science*, 18, 301–306.

CHAPTER 7

A nice paper that describes the characteristics of plants most likely to become locally extinct ('losers') in England (and thus, by implication, of 'winners' also) is Chris Preston's 2000 paper 'Engulfed by suburbia or destroyed by the plough: the ecology of extinction in Middlesex and Cambridgeshire', *Watsonia*, 23, 59–81. My paper showing that successful, increasing plant species all tend to be ecologically rather similar, irrespective of native or alien status, is K. Thompson, J.G. Hodgson and T.C.G. Rich (1995) 'Native and alien invasive plants: more of the same?', *Ecography*, 18, 390–402. Other papers showing much the same thing include S.J. Meiners (2007). Native and exotic plant species exhibit similar population dynamics during succession', *Ecology*, 88, 1098–1104; M.R. Leishman, V.P. Thomson and J. Cooke (2010) 'Native and exotic invasive plants have fundamentally similar carbon capture strategies', *Journal of Ecology,* 98, 28–42; and W. Dawson, M. van Kleunen and M. Fischer (2012) 'Common and rare plant species respond differently to fertilization and competition, whether they are alien or native', *Ecology Letters,* 15, 873–880.

The problems involved in trying to test the enemy release hypothesis are discussed in R.I. Colautti et al. (2004) 'Is invasion success explained by the enemy release hypothesis?', *Ecology Letters,* 7, 721–733. A specific animal example, where what appears to be evidence of enemy release turns out on closer examination not to be, is described by R.I. Colautti et al. (2005) 'Realized vs apparent reduction in enemies of the European starling', *Biological Invasions,* 7, 723–732. Y.J. Chun, M. van Kleunen and W. Dawson (2010) 'The role of enemy release, tolerance and resistance in plant invasions: linking damage to performance', *Ecology Letters,* 13, 937–946, showed that not only do alien plants not always experience enemy release, but it doesn't necessarily do them any good if they do. C. Thebaud and D. Simberloff (2001) 'Are plants really larger in their introduced ranges?', *American Naturalist,* 157, 231–236, showed that European plants introduced to California or the Carolinas, or in the opposite direction, are not generally larger where they are introduced than where they are native.

The review that found propagule pressure is not often examined as a possible cause of invasion success, but that when it is it is almost always found to be important, is R.I. Colautti, I.A. Grigorovich and H.J. MacIsaac (2006) 'Propagule pressure: a null model for biological invasions', *Biological Invasions,* 8, 1023–1037. The paper that documents the key role played by propagule pressure in the success of *Rhododendron ponticum* in Britain is K. Dehnen-Schmutz and M. Williamson (2006) '*Rhododendron ponticum* in Britain and Ireland: social, economic and ecological factors in its successful invasion', *Environment and History,* 12, 325–350, while the paper that does the same for birds introduced to New Zealand is T.M. Blackburn et al. (2011) 'Passerine introductions to New Zealand support a positive effect of propagule pressure on establishment success', *Biodiversity and Conservation,* 20, 2189–2199.

CHAPTER 8

E.S. Zavaleta, R.J. Hobbs and H.A. Mooney (2001) 'Viewing invasive species removal in a whole-ecosystem context', *Trends in Ecology & Evolution,* 16, 454–459, describes several examples of the unintended consequences of the control or eradication of alien species. One specific example is the effect of removing pigs and goats from Sarigan Island in the Northern Mariana Islands, described in C.C. Kessler (2002) 'Eradication of feral goats and pigs and consequences for other biota on Sarigan Island, Commonwealth of the Northern Mariana Islands',

Occasional Papers of the IUCN Species Survival Commission, 27, 132–140. Another is the release of the invasive plant *Senecio madagascariensis* in Hawaii following the removal of goats, described by J.R. Kellner et al. (2011) 'Remote analysis of biological invasion and the impact of enemy release', *Ecological Applications,* 21, 2094–2104.

The importance of the order in which invasive species are removed is illustrated by P.W. Collins, B.C. Latta and G.W. Roemer (2009) 'Does the order of invasive species removal matter? The case of the eagle and the pig', *PLOS ONE,* 4, e7005. The largely unsuccessful 'Control of Invasive Species in the Galapagos Archipelago' programme is described by M.R. Gardener, R. Atkinson and J.L. Renteria (2010) 'Eradications and people: lessons from the plant eradication program in Galapagos', *Restoration Ecology,* 18, 20–29, while the difficulty of eradicating anything other than small populations is described in T. Pluess et al. (2012) 'When are eradication campaigns successful? A test of common assumptions', *Biological Invasions,* 14, 1365–1378. M.R. Gardener et al. (2010) 'Evaluating the long-term project to eradicate the rangeland weed *Martynia annua* L.: linking community with conservation', *Rangeland Journal,* 32, 407–417, tells the story of the devil's claw eradication programme, while T.C. Skurski, B.D. Maxwell and L.J. Rew (2013) 'Ecological tradeoffs in non-native plant management', *Biological Conservation,* 159, 292–302, reports that spotted knapweed, despite being the target of intensive control efforts, has rather little effect on native plants.

The value of invasive Asian honeysuckles for native American birds is shown by J.M. Gleditsch and T.A. Carlo (2011) 'Fruit quantity of invasive shrubs predicts the abundance of common native avian frugivores in central Pennsylvania', *Diversity and Distributions,* 17, 244–253, while the value of alien species as pollinators in New Zealand is demonstrated by D.E. Pattemore and D.S. Wilcove (2012) 'Invasive rats and recent colonist birds partially compensate for the loss of endemic New Zealand pollinators', *Proceedings of the Royal Society B: Biological Sciences,* 279, 1597–1605. The value of working with, rather than against, alien plants in the Don Valley Brick Works in Ontario is described in J. Foster and L.A. Sandberg (2004) 'Friends or foe? Invasive species and public green space in Toronto', *Geographical Review,* 94, 178–198. The rise and fall of Asiatic sand sedge as a sand stabiliser in New Jersey is recounted in L.S. Wootton et al. (2005) 'When invasive species have benefits as well as costs: managing *Carex kobomugi* (Asiatic sand sedge) in New Jersey's coastal dunes', *Biological Invasions,* 7, 1017–1027.

Following accidental introduction to the USA, the Argentinean moth *Cactoblastis cactorum* may or may not eventually prove to be a serious threat to cacti there. The latest position is described in H. Jezorek, A.J. Baker and P. Stiling (2012) 'Effects of *Cactoblastis cactorum* on the survival and growth of North American *Opuntia'*, *Biological Invasions,* 14,

2355–2367. The attempted control of the giant African snail *Achatina* by a predatory snail from Florida, *Euglandina rosea*, and the subsequent very bad behaviour of the latter, is described at length by L. Civeyrel and D. Simberloff (1996) 'A tale of two snails: is the cure worse than the disease?' *Biodiversity and Conservation*, 5, 1231–1252.

The extraordinarily convoluted tale of spotted knapweed and its interaction with the European flies introduced to control it, native deer mice, great horned owls, and ultimately native plants, is told in D.E. Pearson and R.M. Callaway (2008) 'Weed-biocontrol insects reduce native-plant recruitment through second-order apparent competition', *Ecological Applications*, 18, 1489–1500.

The quote from the UK Department for Environment, Food and Rural Affairs ('There is little value …') is from DEFRA (2009) 'Government response to the public consultation: review of Schedule 9 to the Wildlife and Countryside Act 1981 and the Ban on Sale of Certain Non-native Species. Part 1: Schedule 9 Amendments', while the attempt to define 'wild' is from DEFRA (2010) 'Guidance on section 14 of the Wildlife and Countryside Act, 1981'. Both documents are available online. Quotes from the Wildlife and Natural Environment (Scotland) Act 2011 can be found at www.scotland. gov.uk/Topics/Environment/Wildlife-Habitats/InvasiveSpecies/legislation.

The attempt by the Parks Department to remove *Eucalyptus* from Angel Island State Park in California is described by W.E. Westman (1990) 'Park management of exotic plant species: problems and issues', *Conservation Biology*, 4, 251–260. The quote from Westman is from the same source.

The figures for the success rate of species introduced as biological control agents are from R.H. Messing and M.G. Wright (2006) 'Biological control of invasive species: solution or pollution?', *Frontiers in Ecology and the Environment*, 4, 132–140.

CHAPTER 9

The idea that an 'agricultural' response to alien species is likely only to encourage more invasion is explored in R.G. Smith et al. (2006) 'Lessons from agriculture may improve the management of invasive plants in wildland systems', *Frontiers in Ecology and the Environment*, 4, 428–434, while the Kenyan guava story is told in D.G. Berens et al. (2008) 'Exotic guavas are foci of forest regeneration in Kenyan farmland', *Biotropica*, 40, 104–112. The development of pioneer communities of invasive species into (mostly) native forest is described by A.E. Lugo (2004) 'The outcome of alien tree invasions in Puerto Rico', *Frontiers in Ecology and the Environment*, 2, 265–273. The successional pathway of gorse communities in New Zealand is described by J.J. Sullivan, P.A. Williams and S.M. Timmins (2007) 'Secondary forest succession differs through naturalised gorse and native kanuka near Wellington and Nelson', *New Zealand Journal of Ecology*, 31, 22–38, and the eventual disappearance of alien species from forests in Ohio is described by G.R. Matlack and J.R. Schaub (2011) 'Long-term persistence and spatial assortment of nonnative plant species in second-growth forests', *Ecography*, 34, 649–658.

The following three papers describe the global expansion of the Argentine ant and its collapse in New Zealand: V. Vogel et al. (2010) 'The worldwide expansion of the Argentine ant', *Diversity and Distributions*, 16, 170–186; D.F. Ward et al. (2010) 'Twenty years of Argentine ants in New Zealand: past research and future priorities for applied management', *New Zealand Entomologist*, 33, 68–78; M. Cooling et al. (2012) 'The widespread collapse of an invasive species: Argentine ants (*Linepithema humile*) in New Zealand', *Biology Letters*, 8, 430–433.

The rise and fall of the exotic grass *Melinis minutiflora* in Hawaii is described in S.G. Yelenik and C.M. D'Antonio (2013) 'Self-reinforcing impacts of plant invasions change over time', *Nature*, 503, 517–520. A striking feature of this and earlier examples is, however bad things look in the short term, it's surprising how often a little patience is rewarded by a quite dramatic improvement. Another example concerns the removal in 1998 of pigs and goats from Sarigan Island in the Northern Mariana Islands, which I mentioned in the previous chapter. The immediate consequence was a blanket of an invasive vine that looked (as these things often do) like it would last forever. But by 2006 the vine had stabilised, native trees and animal wildlife were both rapidly increasing, and the prognosis is good, reports C.C. Kessler (2011) 'Invasive species removal and ecosystem recovery in the Mariana Islands; challenges and outcomes on Sarigan and Anatahan', *Occasional Papers of the IUCN Species Survival Commission*, 42, 320–324.

The mid-Holocene decline of hemlock in the USA is described in N. Bhiry and L. Filion (1996) 'Mid-Holocene hemlock decline in eastern North America linked with phytophagous insect activity', *Quaternary Research*, 45, 312–320.

The *Bombus subterraneus* story is told in G.C. Lye, O. Lepais and D. Goulson (2011) 'Reconstructing demographic events from population genetic data: the introduction of bumblebees to New Zealand', *Molecular Ecology*, 20, 2888–2900, while the rapid evolution of introduced plants in Australia is described by J.M. Buswell, A.T. Moles and S. Hartley (2011) 'Is rapid evolution common in introduced plant species?', *Journal of Ecology*, 99, 214–224.

The hybridisation of wild and crop radishes in California is described by S.G. Hegde et al. (2006) 'The evolution of California's wild radish has resulted in the extinction of its progenitors', *Evolution,* 60, 1187–1197, and evolution of native Australian snakes in response to the invasive cane toad by B.L. Phillips and R. Shine (2004) 'Adapting to an invasive species: toxic cane toads induce morphological change in Australian snakes', *Proceedings of the National Academy of Sciences of the United States of America,* 101, 17150–17155.

Evolution of native American fruit flies in response to introduced apples and honeysuckles respectively are described in K.E. Filchak, J.B. Roethele and J.L. Feder (2000) 'Natural selection and sympatric divergence in the apple maggot *Rhagoletis pomonella',* *Nature,* 407, 739–742, and D. Schwarz et al. (2005) 'Host shift to an invasive plant triggers rapid animal hybrid speciation', *Nature,* 436, 546–549. The evolutionary impact of Russian knapweed on native grasses, and of fish on native water fleas, is described in B.A. Mealor and A.L. Hild (2007) 'Post-invasion evolution of native plant populations: a test of biological resilience', *Oikos,* 116, 1493–1500, and D.L. Fisk et al. (2007) 'Rapid evolution in response to introduced predators I: rates and patterns of morphological and life-history trait divergence', *BMC Evolutionary Biology,* 7, 22.

The remarkable soapberry bug story is told in two papers by Scott Carroll: S.P. Carroll and C. Boyd (1992) 'Host race radiation in the soapberry bug: natural history with the history', *Evolution,* 46, 1052–1069, and S.P. Carroll et al. (2005) 'And the beak shall inherit – evolution in response to invasion', *Ecology Letters,* 8, 944–951. For the latest in the story, which shows that the Australian soapberry bug has taken radically different genetic routes to feeding on *two* introduced balloon vines, see J.A. Andres et al. (2013) 'Hybridization and adaptation to introduced balloon vines in an Australian soapberry bug', *Molecular Ecology,* 22, 6116–6130.

The discovery that evidence for enemy release declines rapidly with time since introduction is reported in C.V. Hawkes (2007) 'Are invaders moving targets? The generality and persistence of advantages in size, reproduction, and enemy release in invasive plant species with time since introduction', *American Naturalist,* 170, 832–843.

CHAPTER 10

The remarkable story of the transformation of the harlequin ladybird into a world-beating superbug is described by B. Facon et al. (2011) 'Inbreeding depression is purged in the invasive insect *Harmonia axyridis',* *Current Biology,* 21, 424–427.

The rate of acquisition of new arthropods by Hawaii is from R.H. Messing and M.G. Wright (2006) 'Biological control of invasive species: solution or pollution?', *Frontiers in Ecology and the Environment,* 4, 132–140, while the story of *Aphalara itadori* is told by R.H. Shaw,

NOTES

S. Bryner and R. Tanner (2009) 'The life history and host range of the Japanese knotweed psyllid, Aphalara itadori Shinji: Potentially the first classical biological weed control agent for the European Union', *Biological Control*, 49, 105–113.

The latest report on the likely impact of New Zealand flatworm on UK earthworms can be found in A.K. Murchie and A.W. Gordon (2013) 'The impact of the "New Zealand flatworm", *Arthurdendyus triangulatus*, on earthworm populations in the field', *Biological Invasions*, 15, 569–586. Not all earthworms are affected, and *Lumbricus terrestris* is most at risk. The discovery of *Phagocata woodworthi* in Loch Ness is described by T.B. Reynoldson, B.D. Smith and P.S. Maitland (1981) 'A species of North-American triclad (Paludicola, Turbellaria) new to Britain found in Loch Ness, Scotland', *Journal of Zoology*, 193, 531–539.

The importance of cultivation in the spread of various species of Iridaceae around the world is highlighted by M. Van Kleunen, S.D. Johnson and M. Fischer (2007) 'Predicting naturalization of southern African Iridaceae in other regions', *Journal of Applied Ecology*, 44, 594–603, and of the ornamental plant trade in general by K. Dehnen-Schmutz et al. (2007) 'A century of the ornamental plant trade and its impact on invasion success', *Diversity and Distributions*, 13, 527–534. The speed of climate change has been worked out by S.R. Loarie et al. (2009) 'The velocity of climate change', *Nature*, 462, 1052–1055, and how cultivation helps plants to keep one step ahead is described by S. Van der Veken et al. (2008) 'Garden plants get a head start on climate change', *Frontiers in Ecology and the Environment*, 6, 212–216. The bugs on the plants from the South African Chelsea exhibit are described in *Royal Horticultural Society Science News*, 9, October 2011.

The view that translocation of species is wrong in principle is expressed by A. Ricciardi and D. Simberloff (2009) 'Assisted colonization is not a viable conservation strategy', *Trends in Ecology & Evolution*, 24, 248–253. The opposing view is forcefully put by Chris Thomas in his 2011 paper, 'Translocation of species, climate change, and the end of trying to recreate past ecological communities', *Trends in Ecology & Evolution*, 26, 216–221. This provoked a furious response from M. Vila and P.E. Hulme (2011) 'Jurassic Park? No thanks', *Trends in Ecology & Evolution*, 26, 496–497, and an equally robust reply from Chris Thomas: 'Anthropocene Park? No alternative', *Trends in Ecology & Evolution*, 26, 497–498.

The research programme looking into the possible future translocation of the threatened Siberian primrose (*Primula nutans* var. *jokelae*) is described in the January 2012 issue of the journal of Botanic Gardens Conservation International, available online at www.bgci.org/resources/article/0706/.

CHAPTER 11

The quote in the first line of this chapter is from J. Firn et al. (2011) 'Abundance of introduced species at home predicts abundance away in herbaceous communities', *Ecology Letters*, 14, 274–281. The meta-analysis of the impacts of introduced plants is M. Vila et al. (2011) 'Ecological impacts of invasive alien plants: a meta-analysis of their effects on species, communities and ecosystems', *Ecology Letters*, 14, 702–708, while the recent study of the biases inherent in the alien plants we choose to study is P.E. Hulme et al. (2013) 'Bias and error in understanding plant invasion impacts', *Trends in Ecology & Evolution*, 28, 212–218.

The study of the ecosystem services provided by Hawaiian forests invaded by alien plants is J. Mascaro, R.F. Hughes and S.A. Schnitzer (2012) 'Novel forests maintain ecosystem processes after the decline of native tree species', *Ecological Monographs*, 82, 221–228.

The global costs of alien species are tallied in D. Pimentel (2011) *Biological Invasions: Economic and Environmental Costs of Alien Plant, Animal, and Microbe Species*, 2nd edition, Boca Raton, CA: CRC Press. Mark Davis's 2009 book *Invasion Biology* is published by Oxford University Press. The brief item about native plants and water supplies is S. Armstrong (1995) 'Rare plants protect Cape's water supplies', *New Scientist*, 145, 8.

Rachel Carson's *Silent Spring* (1962) has been continuously in print since its first appearance.

An academic study of the mechanisms of competitive ability in plants, used entirely wrongly to discredit purple loosestrife, is C.L. Gaudet and P.A. Keddy (1988) 'A comparative approach to predicting competitive ability from plant traits', *Nature*, 334, 242–243. The source for the 'costs' of purple loosestrife is http://refugeassociation.org/advocacy/refuge-issues/invasive-species/purple-loosestrife/, whereas the Reggie Thackston quote is from www.sherpaguides.com/georgia/fire_forest/creature_feature/.

The revisionist view of the Nile perch is from R.E. Gozlan (2008) 'Introduction of non-native freshwater fish: is it all bad?', *Fish and Fisheries*, 9, 106–115, while the story of the recovery of otters and the decline of mink in the UK is told by R.A. McDonald, K. O'Hara and D.J. Morrish (2007) 'Decline of invasive alien mink (*Mustela vison*) is concurrent with recovery of native otters (Lutra lutra)', *Diversity and Distributions*, 13, 92–98.

The attempt to persuade us that numerous bird species introduced to Europe have big enough effects to require control is S. Kumschick and W. Nentwig (2010) 'Some alien birds have as severe an impact as the most effectual alien mammals in Europe', *Biological Conservation*, 143, 2757–2762, while the demolition of that argument can be found in D. Strubbe, A. Shwartz and F. Chiron (2011) 'Concerns regarding the scientific evidence informing impact risk assessment and management recommendations for invasive birds', *Biological Conservation*, 144, 2112–2118. The reply from Kumschick and Nentwig, containing the remarkable couple of sentences quoted, comes from 'Response to Strubbe et al. (2011): impact scoring of invasive birds is justified', *Biological Conservation*, 144, 2747. The last word on the subject is D. Strubbe, F. Chiron and A. Shwartz (2011) 'Reply to Kumschick and Nentwig (2010, 2011): promoting a robust cost–benefit approach for conducting impact risk assessments of invasive species', *Biological Conservation*, 144, 2748–2748, in which they 'disagree with [the] claim that due to the lack of studies, impacts are more likely to be underestimated than overestimated', and also observe that 'In contrast, several studies have indicated that a bias against non-natives has resulted in an emphasis on documenting negative effects.'

The quote about predation of seabirds by storks, herons and similar birds is from A.J. Williams and V.L. Ward (2006) 'Sacred Ibis and gray heron predation of Cape cormorant eggs and chicks; and a review of ciconiiform birds as seabird predators', *Waterbirds*, 29, 321–327. The two studies that show little effect of sparrowhawks and magpies on songbirds in the UK are D.L. Thomson et al. (1998) 'The widespread declines of songbirds in rural Britain do not correlate with the spread of their avian predators', *Proceedings of the Royal Society of London Series B: Biological Sciences*, 265, 2057–2062, and S.E. Newson et al. (2010) 'Population change of avian predators and grey squirrels in England: is there evidence for an impact on avian prey populations?', *Journal of Applied Ecology*, 47, 244–252.

The study of oak savannah in British Columbia is A.S. MacDougall and R. Turkington (2005) 'Are invasive species the drivers or passengers of change in degraded ecosystems?', *Ecology*, 86, 42–55.

The paper that explains how we inappropriately assess risks is D. Gilbert (2011) 'Buried by bad decisions', *Nature*, 474, 275–277, while the quote from the *Independent* newspaper can be found at www.independent.co.uk/news/british-creatures-fight-for-survival-in-rivers-and-meadows-as-aliens-stage-invasion-of-the-wild-1361423.html.

The effects of native predators on salmon in the American Pacific Northwest are described in M.P. Carey et al. (2012) 'Native invaders – challenges for science, management, policy, and society', *Frontiers in Ecology and the Environment*, 10, 373–381.

Many papers report the (possible) causes and consequences of western juniper expansion in the USA – for example, P.T. Soule, P.A. Knapp and H.D. Grissino-Mayer (2004) 'Human agency, environmental drivers, and western juniper establishment during the late holocene', *Ecological Applications*, 14, 96–112. Similarly, there are many studies documenting the rapid expansion of black vultures in the USA, such as B.F. Blackwell et al. (2007) 'Demographics of black vultures in North Carolina', *Journal of Wildlife Management*, 71, 1976–1979.

The paper suggesting that native species are only rarely invasive is D. Simberloff et al. (2012) 'The natives are restless, but not often and mostly when disturbed', *Ecology*, 93, 598–607.

CHAPTER 12

Our 2011 paper is M. Davis et al. (2011) 'Don't judge species on their origins', *Nature*, 474, 153–154, while the reply is D. Simberloff et al. (2011) 'Non-natives: 141 scientists object', *Nature*, 475, 36.

The first Simberloff quote can be found at www.issues.org/13.4/schmit.htm, and the second at www.gcrio.org/CONSEQUENCES/vol2no2/article2.html.

The quote from the survey of invasion biologists is from A.M. Young and B.M.H. Larson (2011) 'Clarifying debates in invasion biology: a survey of invasion biologists', *Environmental Research*, 111, 893–898. Note that neither 'unnaturalness' nor 'undesirability' are measurable quantities – they are at best opinions, or maybe just beliefs.

There are far too many sources to mention for the idea that introduced species cannot contribute to biodiversity, but here's one: O.E. Sala et al. (2000) 'Biodiversity – global biodiversity scenarios for the year 2100', *Science*, 287, 1770–1774.

Mark Sagoff's penetrating 2005 analysis of our hopelessly irrational attitude to alien species is 'Do non-native species threaten the natural environment?', *Journal of Agricultural & Environmental Ethics*, 18, 215–236.

The devil's claw eradication quote is from M.R. Gardener et al. (2010) 'Evaluating the long-term project to eradicate the rangeland weed *Martynia annua* L.: linking community with conservation', *Rangeland Journal*, 32, 407–417, and the quote 'the public does not readily ...' is from G.W. Witmer et al. (2007) 'Management of invasive vertebrates in the United States: an overview', in *Managing Vertebrate Invasive Species: Proceedings of an International Symposium* (ed. G.W. Witmer, W.C. Pitt and K.A. Fagerstone), Fort Collins, CO: USDA/APHIS Wildlife Services, National Wildlife Research Center.

South Africa's alien plant control strategy was assessed by B.W. van Wilgen et al. (2012) 'An assessment of the effectiveness of a large, national-scale invasive alien plant control strategy in South Africa', *Biological Conservation*, 148, 28–38.

For more on the pointlessness of trying to recreate a pristine past, the value of assisted migration, and much more besides, see Emma Marris's 2011 book *Rambunctious Garden – Saving Nature in a Post-Wild World*, New York: Bloomsbury.

INDEX

Figures in *bold italics* indicate captions.

brown tree snake (*Boiga irregularis*)
49, 50–54, *53*, 80, 179, 200–201,
202, 220
buckwheat 176
Bugwood Wiki: Invasipedia 72–74
bumblebees 17–18, 68, 160–161
bunchgrasses 163
butterflies 150, 186

C

Cactoblastis cactorum 143–144
California 21, 58, 60, 135
  Department of Parks and
    Recreation 149–150
camelids 1, 13
camels
  Bactrian 1, 2
  diversity in South America 2, 5
  dromedaries 1, 2
  evolution in North America 1, 2
  extinction in North America 2
  the largest (*Titanotylopus*) 1
  US Army pack animals 26–27
Canada 12, 108
Canada goose 204
Canadian pondweed 67
cane toad (*Bufo marinus*) 162–163,
220, *221*
canids 12
Cape Town 195
capercaillie 34
Cardinale, Bradley 105
Carey, Michael 211, 213
caribou 11
Carroll, Scott 164
carrots 24, 157
Carson, Rachel: *Silent Spring* 50,
153, 198–199, 200
Caspian tern 211, 212
cassowary 10

cats 3, 37, 51, 80, 92, 132
cattle 71, 130, 137
cattle egret 205
Caucasian wingnut tree (*Pterocarya
fraxinifolia*) 187–188
central American land bridge 12
central Asia 1, 2
Centre for Agricultural
  Bioscience International (CABI)
  176
chaffinch 206
chameleon 3
cheatgrass (*Bromus tectorum*) 138
Chelsea Flower Show (2011) 181
chestnut 159
chimpanzees 22, 26, 27
Christmas Island 20
Chusan palm (*Trachycarpus
fortunei*) 180
cichlids 202
City of Toronto Parks and
  Recreation Department 139
clams 56–57
climate change 48, 149, 159, 182
  adapting to 185
  anthropogenic 6
  and extinction of species 14
  growing threat of 179, 183, 186
  and invasive species 8
  and movement of ecosystems 36
climate warming 116, 180
climatic relicts 14–15
cockroaches 26
Cocos Island 51–52
collared dove 18, 35–36, *36*, 207
colonisation 7
  by black bean aphids 18–19
  by collared doves 18, 35
  by purple loosestrife 61
  by tree bumblebees 17–18, 35
  and extinction 108

INDEX

Printed in the USA
CPSIA information can be obtained
at www.ICGtesting.com
JSHW022216140824
68134JS00018B/1096

9 781771 640961